# SCIENTIFIC ENGLISH
## A Guide for Scientists and Other Professionals

by Robert A. Day

ORYX PRESS
1992

The rare Arabian Oryx is believed to have inspired the myth of the unicorn. This desert antelope became virtually extinct in the early 1960s. At that time several groups of international conservationists arranged to have 9 animals sent to the Phoenix Zoo to be the nucleus of a captive breeding herd. Today the Oryx population is over 800, and nearly 400 have been returned to reserves in the Middle East.

Copyright © 1992 by Robert A. Day
Published by The Oryx Press
4041 North Central at Indian School Road
Phoenix, Arizona 85012-3397

Published simultaneously in Canada

*PE*
*1475*
*.038*
*1992*

Printed and bound in the United States of America

∞ The paper used in this publication meets the minimum requirements of American National Standard for Information Science—Permanence of Paper for Printed Library Materials, ANSI Z39.48, 1984.

**Library of Congress Cataloging-in-Publication Data**
Day, Robert A., 1924–
    Scientific English : a guide for scientists and other
professionals / by Robert A. Day
        p. cm
    Includes index.
    ISBN 0-89774-722-4
    1. English language—Technical English. 2. English language-
-Grammar—1950 3. English language—Rhetoric. 4. Technical
writing. I. Title.
PE1475.D38    1992                           92-9672
808'.0665—dc20                               CIP

To BETTY
Nancy and Joe, Bart and Sarah, Robin and Robert,
Hilary, Hannah, Ian, and Matt

# Contents

# Preface

*It should be our pride to teach ourselves as well as we can always to speak as simply and clearly and unpretentiously as possible, and to avoid like the plague the appearance of possessing knowledge which is too deep to be clearly and simply expressed.*

—Karl Popper

This book is brilliantly (if illegitimately) conceived from this germ of thought: When used simply, English is a simple language.

In the corners of our craniums (or crania if you prefer), we have probably all registered the fact—and it is a fact—that the most profound thoughts ever expressed have usually been expressed in simple language. Plato recognized this fact two millennia ago when he said "Beauty of style and harmony and grace and good rhythm depend on simplicity."

Yes, we know this. But too often we forget. When we forget, we use *long* words in *long* sentences. We start with a simple thought, but then we dress it up in fancy words. As a result, we sometimes lose clarity. At worst, the thought has been overwhelmed by the verbiage; at best, the thought takes second place to the fancy ornamentation surrounding it.

I have spent many years at the interface between science and English as an editor and publisher of scientific books and journals. I have learned two facts that now serve as the thesis for this book: The beauty of science is in the science, not in the language used to describe it. The beauty of English is its ability, when properly used, to express the most complicated concepts in relatively clear words and to point up the beauty of the science. Successful communication in science involves that magic word, *clarity,* a kissing cousin of *simplicity.*

To be simple, writing usually needs to be short; that is good for me because it matches my attention span.

But, you say, English has a massive vocabulary, and English professors have concocted a zillion rules; therefore, it is virtually impossible to write in English with clarity and with confidence.

To you, I say, this book and its simple guidelines just might improve your knowledge of English and your ability to communicate in English. If you want to learn about such esoteric things as the subjunctive mood and the pluperfect tense, do *not* read this book; if you want to write with simple, straightforward accuracy, this book just might help. Although it is aimed directly at those who write articles for publication in primary scientific journals, many of this book's principles may also be useful in other types of writing.

English *is* simple. Consider this: English has a truly massive vocabulary of some 500,000 words, but how many different *kinds* of words does it have? The answer is nine, the nine "parts of speech." You could construct nine pigeonholes, and every one of those 500,000 English words would fit into one (or more) of those pigeonholes. Thus, to get a real feel for the use of English, you do not need to master a half-million words (you *will* need a few thousand, however); instead, you need to look at the nine kinds of words and learn a few simple rules about using them.

Then, you can group words into phrases and clauses. How many phrases and clauses are there? There are essentially four main types of phrases and two types of clauses. Of course, you need a few definitions and a few rules. Simple rules.

Next, you can combine phrases and clauses into sentences. How many different types? A mere six. Every sentence ever constructed in English is of one of these six types. Relatively simple rules for constructing and punctuating these six types of sentences can be stated.

Finally, knowing how to construct sentences, the basic building blocks of communication, you can go on to paragraphs and then to papers and books and all the rest.

Am I oversimplifying? Perhaps. But I have tried in this book to simplify English. In the past, too many grammarians have established far too many arcane "rules" about the use of our language. Some of these people believed (or pretended?) that English can be used precisely only by the literati. Common folk (scientists, for example) couldn't be expected to master the profusion of arbitrary rules that were supposed to relate to English. When grammarians dangled their rules, many of us got splitting headaches from worry about dangling participles and split infinitives.

I have good news. You *may* split infinitives. In fact, you may, on occasion, violate every one of the "rules" dreamed up by generations of grammatical fussbudgets. (When I was young, I was told to always obey my superiors. I said that I would, if I ever found any.) The obvious purpose of grammatical rules is to facilitate clear communication. When rules of grammar do not serve this purpose, they should be disregarded.

Remember only one main rule. A sentence is a unit of *thought.* If you write it down, it becomes a unit of expression. The simple truth is this: If you can think logically, you can write logically. A logical sentence is a good sentence. Now let's turn that around: A good sentence is a logical sentence. So, just possibly, you might learn to *think* more clearly if you learn to *write* more clearly. Perhaps this book will help.

---

Clear communication, which is the prime objective of scientific reporting, may be achieved by presenting ideas in an orderly manner and by expressing oneself smoothly and precisely. By developing ideas clearly and logically, you invite readers to read, encourage them to continue, and make their task agreeable by leading them smoothly from thought to thought.

—*Publication Manual* of the American Psychological Association

# Acknowledgments

A number of very talented people read the manuscript of this book and generously offered critical comments. Because of their help and advice, this book is very much better than it otherwise would be. So, I offer sincere thanks to the following colleagues: Deborah C. Andrews, R. John Brockmann, Barton D. Day, Betty J. Day, Robin A. Day, Sarah F. Day, Joan DelFattore, Barbara Gastel, Jay Halio, Marcia Peoples Halio, Ronald E. Martin, Evelyn S. Myers, Joseph Sakaduski, Nancy Sakaduski, Rivers Singleton, Jr., David W. Smith, Robert Snyder, and Harold B. White III.

# Chapter 1
# Principles of Scientific Writing

*It is impossible to dissociate language from science or science from language, because every natural science always involves three things: the sequence of phenomena on which the science is based; the abstract concepts which call these phenomena to mind; and the words in which the concepts are expressed. To call forth a concept a word is needed; to portray a phenomenon, a concept is needed. All three mirror one and the same reality.*

—Antoine Laurent Lavoisier

## KINDS OF WRITING

Writing can be used in many different ways to express ideas. Thus, there are many kinds of writing. One style of writing might be appropriate for one purpose but totally inappropriate for another. Many of the "writing" courses in our colleges and universities are appropriate for teaching creative writing. In some ways, however, they are inappropriate for teaching the principles of scientific and technical writing.

I would not for a moment denigrate creative writing. Life would be bleak indeed without the grand heritage provided by our poets, novelists, dramatists, and essayists. However, there is a world of difference between creative writing and scientific writing. The one deals primarily with feeling, emotion, opinion, and persuasion. The other emphasizes the dispassionate, factual recording of the results of scientific investigations. The one uses language of extraordinary beauty and complexity, with fascinating metaphors and other figures of speech; scientific writing uses (or should use) prosaic words of certain meaning, organized simply into precise phrases, clauses, sentences, and paragraphs.

Does this mean that scientific writing never expresses "feeling, emotion, opinion, and persuasion"? Of course not. The practice of science is engaged in by people (scientists) who have the same complement of good and bad points (did someone say "egos"?) as the rest of us. Thus, I admit that the above paragraph is too idealistic, because scientific writing is often, sometimes heavily, infested with persuasion, opinion, etc. But isn't the ideal of "dispassionate, factual recording of the results of scientific investigations" worth striving for? It is the thesis of this book that it is.

Does this mean that scientific writing must be dull? Not necessarily. Dullness can result when writers give up the pretty ornaments of creative writing; however, if clarity is increased, the reader may enjoy the comprehension and not notice the loss of ornaments. In writing, as in the making of jewelry, true elegance often results from the simple rather than the ornate.

A fundamental difference between creative writing and scientific writing was well stated by the novelist John Fowles: "For what good science tries to eliminate, good art seeks to provoke—mystery, which is lethal to the one, and vital to the other."

Scientific writing, the subject of this book, is not the same as "science writing." Both are related, of course, because the subject matter of both is science. However, an important distinction to keep in mind is that "scientific writing" is written by scientists for an audience of scientists, whereas science writing is written (sometimes by scientists, sometimes by journalists) for an audience of nonscientists (or scientists reading outside their own narrow discipline). Thus, the vocabulary, tone, and complexity of these two types of writing differ.

Further, scientists must write in different ways for different purposes, using one language in their research papers, a slightly different language in their review papers, and a very different language in communications directed to lay audiences. The emphasis in this book is on the language of research papers, but the guidelines presented and the examples used may be of use in many types of writing.

## GUIDING PRINCIPLES

The principles of scientific writing derive from the purposes of scientific writing. The basic purposes of scientific publications are (1) to *record* (the archival function of our research journals); (2) to *inform* peers; and (3) to *educate* the next generation of scientists.

The archival use of journals is very significant, because a journal in a library can be consulted by a great variety of readers for any number of reasons over a long time period. It is primarily this potential audience that scientists should write for. Jargon and unexplained abbreviations, although understandable to peers, should therefore be avoided. Students also will benefit from such consideration.

I would guess that a great many of the really bad (incomprehensible) papers that appear in our journals are bad because the authors ignored the archival and student audiences for their papers and wrote in the arcane shorthand of laboratory jargon. Moreover, scientists sometimes go out of their way to flaunt this jargon, and they (only the worst of them, fortunately) often combine their jargon with an incredibly verbose style of writing that leads to real confusion.

Too many scientists, and perhaps members of all professions, want to "sound" scholarly. Therefore, they sometimes dress up a simple thought in an outrageous costume. Sometimes, the thread of the idea gets lost along the way, and all we see is the frayed costume.

As for me, I don't want the costume. If I have learned anything from my years of experience in scientific writing, editing, and publishing, it is this: Simplicity of expression is a natural result of profound thought.

---

The theory that scientific discovery is impersonal or, as it is called, objective, has had several evil consequences. One is that the style of describing and publishing the results of scientific research which is fashionable today has been developed to sustain it. One should write, one is told, in the third person, in the passive voice, without betraying conviction or emphasis, without allusion to any concrete or everyday object but with the feeblest indifference and the greatest abstraction. This practice has proved to be so readily acquired that it has now, for a whole generation, been debauching the literary languages of the world. The result has been that science, instead of being a source of strength and honesty is in fact robbing the common speech of these very qualities. For the style itself is neither strong nor honest.

—C. D. Darlington

# Chapter 2
# Style of Scientific
# Writing

*By all means, you should write in your own personal style, but keep in mind that scientific writing is not literary writing. Scientific writing serves a completely different purpose from literary writing, and it must therefore be much more precise.*
      *—The ACS* (American Chemical Society) *Style Guide*

## DEFINITION OF STYLE

The word "style," when applied to writing, pertains not only to writing style, but also to the basic organization of a scientific paper or other publication, the editorial style of the journal or publisher, and the typographical style of the publisher or printer. In short, style defines the *personality* of a publication. Each publication has its own style, its own personality. Well-edited journals have very distinctive styles. Careful writers make it their business to learn the general stylistic conventions used in their field and also the specific style requirements of the particular journal for which they are preparing a manuscript.

## GENERAL STYLE IN SCIENTIFIC WRITING

In a general way, the style of scientific writing is (or should be) distinctive in two principal ways. First and foremost, as already stated, scientific writing should be simple and clear. Its purpose is not to entertain or to paint pretty pictures, but to inform.

But, you say, why should scientific writing be simple, plain, and ordinary? Why can't it be *interesting?* I would argue, and argue strongly, that *good* scientific writing is often beautiful in its elegant simplicity. However, if you are faced with a choice between expressing a thought with a beautiful but complex metaphor or with simple, concrete words, choose the concrete words. There is really only one essential goal in scientific writing: clarity. And short words are not necessarily dull. Lincoln's famous Gettysburg Address contained only 267 words, of which 196 (73%) were one-syllable words.

The second general aspect of style in scientific writing is organization. Of course, all types of writing are "organized"; however, scientific writing is rigidly organized, and each scientific paper is organized *in the same way*. (This is true for the vast majority of research papers, less true for other types of papers produced by scientists.) This type of organization has come to be known by the acronym IMRAD (Introduction, Methods, Results, and Discussion). Inasmuch as a book describing the IMRAD system of organization already exists (Day, R.A., *How to Write and Publish a Scientific Paper,* 3rd ed., Oryx Press, Phoenix, 1988), I shall not expand upon the subject here.

## SPECIFIC STYLE IN SCIENTIFIC WRITING

Style has so many aspects that entire books have been written on the subject. These books (style manuals) are filled with useful information pertaining to specific fields. I strongly recommend that every scientist own at least one style manual. Biologists should own the *CBE Style Manual: Guide for Authors, Editors, and Publishers in the Biological Sciences,* 5th ed., Council of Biology Editors, Inc., Bethesda, MD, 1983. Microbiologists should own, in addition, a copy of the *ASM Style Manual for Journals and Books,* American Society for Microbiology, Washington, DC, 1991. Chemists should own J.S. Dodd's *The ACS Style Guide: A Manual for Authors and Editors,* American Chemical Society, Washington, DC, 1986. Medical and biomedical researchers should own either (or both) the *American Medical Association Manual of Style,* 8th ed., Williams & Wilkins, Baltimore, 1989, or E.J. Huth's *Medical Style & Format: An International Manual for Authors, Editors, and Publishers,* Williams & Wilkins, Baltimore, 1987. Psychologists should own the APA's *Publication Manual,* 3rd ed., American Psychological Association, Washington, DC, 1983. Professionals in *any* field can find a wealth of information in *The Chicago Manual of Style,* 13th ed., University of Chicago Press, Chicago, 1982.

I also recommend that every scientist read the Instructions to Authors of a journal before starting to write a paper, consult the Instructions while writing the paper, and check the paper against the Instructions before submitting it. These Instructions are usually short, but they often contain the highly specific information that defines the editorial and typographical personality of that particular journal as well as the basics of how and where to submit manuscripts. Authors who ignore these specific journal requirements are likely to receive many rejection letters.

## SPELLING AND GRAMMAR

Naturally, proper grammar should be used, and words should be spelled correctly. But correctness may sometimes depend on style. Is it "color" or "colour"? The answer depends on whether the journal uses American English or British English. Is "labeled" or "labelled" correct? Unfortunately, the dictionaries seem to be about evenly split, so both are "correct." However, journals and publishers are likely to choose one or the other, and their copy editors will attempt to invoke this adopted style consistently. Is it "ameba" or "amoeba"? Is it "orthopedics" or "orthopaedics"? The simpler (and newer) spelling seems to be replacing the older usage, but many journals retain the older style.

Does this matter? In individual instances, such choices seem to have little meaning. But, collectively, such considerations may have great importance. For one thing, consistent spelling simplifies things for the reader, especially for the reader whose native language is not English. If the same word is spelled two different ways in one article or in one issue of a journal, a reader is likely to assume that the different spellings somehow mean different things. Confusion, or at least delayed comprehension, can result.

More importantly, slight alterations in spelling sometimes do indeed mean different things. Two words commonly confused by some scientists are "phosphorus" and "phosphorous"? Both are correct, but the meanings are different. The word "phosphorus" is a noun, and it refers to the element phosphorus. The word "phosphorous" is an adjective, not a noun; it refers to compounds containing phosphorus, in particular those with a valence lower than the valence in phospho*ric* compounds.

Are such "slight alterations in spelling" really significant? Yes, for two main reasons. First, even a slight misspelling can cause confusion of meaning. Second, misspellings can foil even the best computer search programs; as scientists increasingly rely on computers to access journal articles, variations in spelling become ever more dangerous.

Stylistic conventions relate not only to spelling but to such things as word choice and capitalization. Which is correct: formalin, Formalin, or formaldehyde? The term formaldehyde is a generic name, and it is normally acceptable. The term Formalin is a registered trademark, and this proprietary status is recognized by the capital F. The use of "formalin" without a capital is incorrect. Finally, "Formalin" should not be cavalierly changed to "formaldehyde" (on the grounds that most journals prefer generic names to proprietary names), because formaldehyde exists as a gas and Formalin is a solution containing a small amount of methanol.

## BY THE NUMBERS

Because numbers are used so often in science, it will be helpful to memorize the stylistic rule that a great many scientific publications follow: Spell out one-digit numbers (one to nine) and use numerals for all larger numbers (10 and up).

> One for the money, two for the show.
>
> All 13 of us went to the lecture.

Now that you have memorized that simple rule, I must unfortunately ask you to memorize the main exceptions.

Spell out any number that starts a sentence.

> Thirteen of us went to the lecture.

Use numerals whenever numbers are followed by units of measure.

> I added 3 ml of distilled water.

In a series, use numerals if any number in the series is 10 or more.

> I did 4 experiments on Monday, 5 on Tuesday, and 11 on Wednesday.

## PRINTING STYLE

The careful author consults the Instructions to Authors *and* a recent issue of a journal before submitting a manuscript to that journal. The author needs answers to such questions as these: What types of headings and subheadings are used? Are footnotes allowed? Which style of literature citation is used? What is the format for tables and figures and their legends? How are chemical and mathematical formulas presented? Get all the answers, and then write with style.

---

It is usually found that only stuffy little men object to what is called "popularization," by which they mean writing with a clarity understandable to one not familiar with the tricks and codes of the cult. We have not known a single great scientist who could not discourse freely and interestingly with a child. Can it be that the haters of clarity have nothing to say, have observed nothing, have no clear picture of even their own fields?

—John Steinbeck and Ed Ricketts

# Chapter 3
# The English Language

*Language is the only instrument of science, and words are but the signs of ideas.*

—Samuel Johnson

## THE BEAUTY OF ENGLISH

Modern science contends with some extremely complicated problems. Hence, the language scientists communicate in must be capable of precise descriptions of complex problems and concepts.

Fortunately, scientists have such a language in English. The skilled user of English has a rich supply of words to describe and differentiate the finest gradations of meaning. We can even play games. For example, the word "stand" has many different meanings: two common usages are "to rise to an erect position" and "to tolerate." Thus, we can describe a person who doesn't like desk work as one who might say "I can't stand sitting." A strange concept, perhaps, but probably no stranger than the classic "I think I will sit out this dance." The remarkable thing, of course, is that the English language has this tremendous array of words that can be constructed into phrases, clauses, and sentences in a seemingly inexhaustible variety of ways if we play by the rules of logic.

## ENGLISH—THE INTERNATIONAL LANGUAGE

It has often been said that science is international. Now it can be said that English is international. For scientists, especially, English is virtually the *only* language.

Look at what has happened in the field of microbiology, for example. For many years, the principal language of this science was German, and the leading journal was the renowned *Zentralblatt für Bakteriologie.* This distinguished journal is still published, and the title remains the same. But the articles are in English.

The *Journal of Antibiotics,* published in Tokyo, is perhaps the most important journal in the world dealing with this subject. Every word is in English.

For French, perhaps the denouement (one of many useful words that came into English from French) to its use as a language of science occurred in 1989, when the January issue of the famous *Annales de l'Institut Pasteur* was published under a new name, *Research in Immunology.* Every article was in English, as were the Instructions to Authors, the book reviews, and even the small-print subscription information. An accompanying editorial had this to say:

> Times have changed . . . biological sciences, and immunology in particular, have evolved, with an enormous increase in the volume of research work performed and very distinct requirements in terms of scientific communication . . . . All journals must now find their roots in the international community at large and interact with a wide network of scientists and institutions. Such will be the case for "Research in Immunology", which will now replace the one hundred and one year old "Annales".

## UNIVERSALITY OF ENGLISH

English is not just the international language of science. Increasingly, English is becoming the international language of business and of the computer.

In late 1986, Public Television presented a six-part series (since published in book form) titled "The Story of English." In the first scene of the first episode, viewers saw and heard the conversation between an airplane pilot and a control tower operator while the plane was being brought in for a landing. Every word was in English. Eventually, the narrator broke in to inform the audience that they were witnessing a pilot-tower conversation during a flight originating and ending in Italy. The tower, of course, must handle both domestic and international flights, and use of a single language is a necessity.

The world is getting smaller indeed, and English is becoming the international language. Any final resistance to this trend will be wiped out by the economics of the computer. Computer software and documentation are expensive, and duplication in different languages is often prohibitive. Thus, as computers become ubiquitous, the Tower of Babel will be replaced by English-speaking control towers.

## RESPONSIBLE USE OF ENGLISH

So, if English is now the international, universal language, what does this mean to scientists? Broadly, it means that scientists must accept the responsibility of using English with precision. All scientists, wherever they are in the world and whatever their native language, must acquire reasonable fluency in English. Except in a few small scientific backwaters, it simply is no longer possible to do science except in English. No longer can a scientist keep up with the literature, or contribute to the literature, without command of the English language. No longer can one depend on colleagues for help in literature searches or for translation of manuscripts. Just as modern scientists must learn the intricacies of increasingly complicated laboratory equipment and experimental protocols, they must also learn to weigh out their words, in English, with the same precision they must use in weighing out reagents in their laboratories.

Scientists (and perhaps scholars in all fields) should learn to use English *simply.* Short simple words—in short, straightforward sentences—usually convey meaning more clearly than do esoteric words and convoluted sentences. This concept is a bit controversial, because the skilled writer, using that wonderful, massive vocabulary we have available in English, can paint word pictures of overwhelming beauty. On the other hand, clarity and meaning can easily be lost in the ornamentation.

The other reason for using simple English in scientific papers, however, should not be controversial. This reason has already been mentioned: We are the *first* generation in the history of the world to use English as the international language of science. However, the majority of this generation grew up using a language *other than* English.

Therefore, as most of our colleagues around the world are struggling to master this strange language we call English, we should all do whatever we can to write clearly and simply. Language, like science itself, can lead to confusion when we are not careful.

---

The chief merit of language is clearness, and we know that nothing detracts so much from this as do unfamiliar terms.

—Galen

# Chapter 4
# Grammar

*Let schoolmasters puzzle their brain,*
*With grammar, and nonsense, and learning;*
*Good liquor, I stoutly maintain,*
*Gives genius a better discerning.*
                                    —Oliver Goldsmith

## RULES OF GRAMMAR

You have no doubt heard a number of the basic rules of English grammar: Do not split infinitives; do not end sentences with a preposition; do not use singular subjects with plural verbs; don't use double negatives, etc. Take my advice: Forget every one of those rules. They are so often wrong or confusing that they have little if any value.

Does that mean English has no rules? Well, there are a few guidelines, a few clues that one can use. More than anything else, there is or should be a philosophy of English. The central tenet of this philosophy I call the "grammar of meaning." After all, the goal of the writer is to convey meaning to the reader. The critical question, then, is whether a sentence conveys clear meaning. If it does, it is a good sentence, no matter how many so-called rules have been broken. If the sentence is not clear, it is a poor sentence, no matter how impeccable the grammar.

## AGREEMENT OF SUBJECTS AND VERBS

If there is any grammatical rule that makes sense, it is the one that says that singular subjects take singular verbs and plural subjects take plural verbs. Most verbs have a form, usually ending in "s," that goes with singular subjects and another form, without the "s," that goes with plural subjects.

She runs.

They run.

You should always keep this rule in mind, and you should follow the rule most of the time. A person who says or writes something like "They was going to the party" is likely to be accused of bad grammar and not invited to the next party.

Do not follow this rule rigidly, however, because, like all grammatical rules, there are many exceptions. In a sense, the rule is indeed valid, once you understand the "grammar of meaning."

A series of experiments (was, were) performed.

A number of experiments (was, were) performed.

Is "was" or "were" correct in the above sentences? In both, the subjects appear to be singular in number. (The subject of the first sentence is "series"; the subject of the second sentence is "number." The word "experiments" in each sentence is the object of a preposition.) In the first sentence, the word "series" (which can be either singular or plural) almost certainly refers to a related group of experiments, considered as a whole or as one group. Therefore, the singular verb form would be correct: A series of experiments *was* performed.

In the second sentence, the word "number" is seemingly singular, but grammar of meaning would tell us that "A number of experiments" is almost certainly more than one (or the author would have said "One experiment"). Thus, we must logically use the plural verb form: A number of experiments *were* performed.

My point is that a milligram of logic is grammatically more important than a kilogram of rules. Consider the following sentences:

A bunch of grapes (is, are) on the table.

A bunch of apples (is, are) on the table.

Following the grammar of meaning, we ask what really is on the table. In the first sentence, we do not have a number of unconnected grapes on the table; we have a "bunch," which (when applied to grapes) means a connected group. Thus, one "bunch" takes a singular verb: A bunch of grapes *is* on the table.

In the second sentence, the word "bunch" applies to apples, which grow as singles and not as connected groups. Here, "bunch" is used, a bit informally, to indicate a group of these single entities. Thus, we have many apples, and logically we must have a plural verb: A bunch of apples *are* on the table.

Scientific and technical writers also use (or should use) this same logic in sentences containing units of measure. Should we say "3 ml was added" or "3 ml were added"? Logic tells us that "3 ml was added" is correct, because *one quantity* was added; whether that quantity is 1 ml or 750 ml is irrelevant. (Rarely, 3 ml of a reagent might be added sequentially, 1 ml at a time; in such an instance, "3 ml were added" would be correct.)

Many errors in agreement relate to a different kind of meaning, and that is simple dictionary meaning.

> If this criteria is met, we will have no problem.
>
> This media lacks glucose.
>
> This data is incomplete.

In these sentences, the writers were evidently unaware that "criteria" is the plural of "criterion," "media" is the plural of "medium," and "data" is the plural of "datum."

Other errors occur because the writer is confused by intervening elements and forgets to match the subject with the verb.

> The use of various acids and other reagents often result in marred surfaces.

The person who wrote this sentence thought the plural verb "result" should follow the plural "acids" and the plural "reagents." However, "acids" and "reagents" are objects of the preposition "of." The subject of the sentence is the singular "use," so the sentence should say "use . . . *results* in marred surfaces."

## SPLIT INFINITIVES

In an earlier book (*How to Write and Publish A Scientific Paper*, 3rd ed., Oryx Press, Phoenix, 1988), I said that "Most of us these days don't worry about things like split infinitives." Oh, but some people *do*. In a journal called *The Observatory*, a reviewer of that book cited that sentence as "unacceptable opinion." Many people seem to find English, and perhaps life itself, difficult in the absence of rigid rules or laws. Nevertheless, there is no valid rule against split infinitives. And those people who insist on unsplitting infinitives often end up with awkward or confusing sentences. Using a split infinitive, I can say

> I fail *to completely understand* rigid rules.

People who abhor split infinitives would place "completely" just before or just after the infinitive.

> I fail completely to understand rigid rules.
>
> I fail to understand completely rigid rules.

Note how the meaning of the original sentence ("completely understand") has been drastically altered in the second sentence (in which the reader assumes that I "fail completely") and also in the third sentence (in which the reader assumes that I am talking about "completely rigid" rules). Thus, in this very common kind of sentence, the only safe place for the adverb ("completely") is smack in the middle of the infinitive phrase ("to completely understand").

## DOUBLE NEGATIVES

The rule is usually expressed as "Don't use no double negatives." Like most rules, this one makes sense if it is used as a general guideline rather than as a rigid rule. Often, two (or more) negatives can be used in the same sentence successfully. The meaning of the following rustic sentence is crystal clear, even though it contains a quintuple negative:

> Ain't nobody around here who knows nuthin' about nuthin' nohow.

On the other hand, a second negative sometimes has the effect of canceling the first, giving the sentence an unintended positive meaning. These we have to watch for. Few of us would have problems with such obvious double negatives as "I ain't got no money." Such constructions sound ungrammatical and they are, although their meaning is clear (and negative). These are easy to spot when obvious negatives (no, not, nor, never) are used. More troublesome are the hidden negatives. For example, "unwell" is a negative of "well."

> She is well.
>
> She is not well.
>
> She is not unwell.

The meaning of the first two sentences is clear. The third sentence, however, contains two negatives, the second canceling the first. (If she is "not unwell," she must be "well.") If she really is "well," the sentence should be stated positively. Other common (and confusing) double negatives are "not infrequently" and "hardly uncontroversial." Some of the worst sentences are those that start out with confusing double negatives, such as "Although it is not unknown for scientists to . . ." In such sentences, comprehension is lost before we even get to the main clause.

> Sometimes people who are lost in a strange city feel uncertain about asking for help. Don't.

This example is not a double negative; it is a single negative. But what is being negated? Are we being told not to get lost, not to feel uncertain, or not to ask for help? A good rule, I think, is to follow the words of the old song: "accentuate the positive; eliminate the negative; and don't mess with Mr. In-Between."

## SYNTAX

Will Rogers said there must be something wrong with "syntax" because it has both "sin" and "tax" in it. Indeed, there can be many things wrong with syntax, which is the branch of grammar dealing with word order. For a sentence to make sense, the words must be presented in a logical sequence. If the words are not in reasonable order, the result can be at least confusing and sometimes ridiculous.

I knew a man with a wooden leg named George.

Few people can recognize participles, dangling or otherwise. However, there is no great need to learn about participles and other arcane niceties of English. What you need to learn is the fundamental principle of syntax: *modifiers should be as close as possible to the words, phrases, or clauses they modify.* This "rule" is nothing more than logic: If words relate to each other, they should be near each other.

Some modifiers really "dangle," in that they have nothing to modify.

While having lunch, the reaction mixture exploded.

In analyzing the data statistically, the *Salmonella typhimurium* infections were indeed rare.

Obviously, "the reaction mixture" was not having lunch. Presumably, a scientist or a student or somebody was having lunch, but the sentence does not yield this information; therefore, the sentence is silly. The second sentence is just as silly because *S. typhimurium* infections can't analyze data (statistically or otherwise). Yet, sentences like this, in which the agent of the action has been omitted, abound in scientific writing.

Single words, usually adverbs, can cause problems if the writer is careless about where such words are inserted in the sentence. The most common offender is the word "only." For example, "only" can be inserted anywhere in the following sentence. However, read the versions of the sentence carefully, and you will note considerable variations in meaning.

Only I hit him in the eye yesterday.

I only hit him in the eye yesterday.

I hit only him in the eye yesterday.

I hit him only in the eye yesterday.

I hit him in only the eye yesterday.

I hit him in the only eye yesterday.

I hit him in the eye only yesterday.

I hit him in the eye yesterday only.

The variations in meaning range from "Only I" to "yesterday only," visiting a one-eyed man ("the only eye") along the way.

The word "only" is not the only word to watch; the syntactical location of the word "just" is just as important.

Just today we visited my aunt.

Today just we visited my aunt.

Today we just visited my aunt.

Today we visited just my aunt.

Today we visited my just aunt. (I also have an "unjust" aunt.)

Consider that there is almost $1,000 worth of difference between the following two sentences:

I almost wrote a check for $1,000.

I wrote a check for almost $1,000.

Finally, don't worry about syntax. It's all right with me if you forget you ever heard of it. But remember logic. If you think and write logically, you will not be guilty of a sentence such as the following:

I went to a town that was 20 miles away on Tuesday.

You might not notice the problem in syntax. The prepositional phrase "on Tuesday" is too far away from the word it modifies ("went"). But you should notice the problem in logic: If the town was "20 miles away on Tuesday," how far away was it on Monday?

## THERE'S THE RUB

There is nothing wrong with a sentence beginning with "There." (See, I just did it.) However, this form of indirect expression is wordy and should rarely be used. The first "There" in the above sentence could be deleted and the sentence reworded to say "Nothing is wrong with a sentence beginning with 'There.'" There are many other sentences that can be improved by avoiding the "There" opening. (Many other sentences can be improved by avoiding the "There" opening.)

---

A farm-country girl enrolled at the state university. In her first term, she made the mistake of associating with one of the college conquistadors. When she went home for the Christmas holidays, she at least wanted to be honest. So she said to her father, "Paw, there's something I have to tell you. I ain't a good girl anymore." Her father replied, "What! Your mother and I scrimped and saved for years so that you could go to that big hot-shot university. Then, after three full months, you come home, and you're still saying ain't."

---

# Chapter 5
# Words

*Long words name little things. All big things have little names, such as life and death, peace and war, or dawn, day, night, love, home. Learn to use little words in a big way. It is hard to do. But they say what you mean. When you don't know what you mean, use big words. They often fool little people.*

—*SSC Booknews,* July 1981

## THE TAXONOMY OF WORDS

Because of its rich vocabulary, the English language can be used to describe both thoughts and things with exquisite precision. Many of these words have unique meanings. Other words have nebulous meanings or mean the same as one or more other words in the language. The unique words, those of certain meaning, are obviously the words of first choice (even if they are long words). More often, however, the choice is between or among words that are essentially synonymous; here, the writer should use the short word or the common word. Usually, the most common word *is* the shortest. However, these choices are seldom easy. The very richness of the English language is daunting to all writers, not just scientists.

## CHOICE OF WORDS

To emphasize the daunting nature of English, I will provide a few examples. The word *oversight* can mean responsibility or the lack of it. The words *valuable* and *invaluable* mean the same thing. A *reckless driver* is not likely to be a *wreckless driver.* Will an *inflammable* substance burn? Yes, because it is *flammable.* Most psychic mediums are ill; a *well medium* is *rare.* (I have a big stake in this viewpoint.)

18

Nonetheless, in spite of the oddities, we can use English reasonably if we try to use the short word, the known word, the word with certain meaning.

The short word is usually obvious. What is the *known* word? Usually, the known word is the common word. But common to whom? One person's jargon may be someone else's everyday speech.

> Two small boys were playing. One said: "I found a condom on the patio."
> The other asked: "What's a patio?"

Choose the right form of the right word and stick with it. In English courses, you were probably taught to vary words for the sake of variety. That is fine for literary writing but not for scientific writing. Keep in mind that every such variation can be confusing, especially to nonnative readers who are still struggling with English. Some particularly troublesome words and expressions are given in Appendix 2.

## THE PARTS OF SPEECH

Fortunately, the taxonomy of English words is relatively simple. If we take the half-million English words and sort them into nine taxonomic pigeonholes, we can learn some simple definitions and rules about these nine categories. Then we can use, confidently and with precision, thousands of the words we have at our disposal. If we once get it clear in our heads what a "verb" is, for example, we can then effectively use many of the verbs available to us in the English language.

These nine pigeonholes are referred to as the "parts of speech": nouns, pronouns, verbs, adjectives, adverbs, conjunctions, prepositions, interjections, and articles. (Some words fit into more than one pigeonhole.) These parts of speech are briefly discussed in the next four chapters.

### Parts of Speech

Three little words you often see
Are ARTICLES, *a, an*, and *the*.
A NOUN's the name of anything;
As *school* or *garden, hoop* or *swing*.
ADJECTIVES tell the kind of noun;
As *great, small, pretty, white*, or *brown*.
Instead of nouns the PRONOUNS stand;
*Her* face, *his* face, *our* arms, *your* hand.
VERBS tell of something being done;
To *read, count, sing, laugh, jump*, or *run*.
How things are done the ADVERBS tell;

As *slowly, quickly, ill,* or *well.*
CONJUNCTIONS join the words together;
As men *and* women, wind *or* weather;
The PREPOSITION stands before
A noun, as *in* or *through* a door.
The INTERJECTION shows surprise;
As *oh! how pretty! ah! how wise!*
The whole are called nine parts of speech,
Which reading, writing, speaking teach.

—Anonymous

# Chapter 6
# Name Words (Nouns and Pronouns)

*Big words can bog down: one may have to read them three or four times to make out what they mean. Small words are the ones we seem to have known from the time we were born, like the hearth fire that warms the house.*

*Short words are bright like sparks that glow in the night, moist like the sea that laps the shore, sharp like the blade of a knife, hot like salt tears that scald the cheek, quick like moths that flit from flame to flame, and terse like the dart and sting of a bee.*

—Celia Wren

## NOUNS

A noun is a word for a person, place, thing, or idea. Having provided that definition, I will now ring a small warning bell. You will find many simple definitions in this book. I prefer to think of them as both simple and correct definitions, but I admit that my goal in writing this book is to provide a *simple* explanation of English usage. By no means is this a *complete* explanation. I am convinced that scientists (and even English professors, for that matter) can write clear, logical sentences by following a few, simple rules. So, when I say a noun is a person, place, thing, or idea, accept this definition without argument; this will help you to recognize and to use effectively thousands of English words. If, at the back of your mind, you cling to the notion that exceptions to this rule exist, that might be wise. Even if my definitions and explanations were far more sophisticated than they are, exceptions would still exist. In fact, I am unaware of *any* rule regarding English usage that does not have exceptions.

## Proper and Common Nouns

There are two types of nouns: proper nouns and common nouns. A proper noun is a *specific* person, place, thing, or idea. Proper nouns include specific persons (Robert A. Day, George Washington), places (Chicago, Death Valley), things (Grant's Tomb, Empire State Building), and ideas (Methodism, Marxism).

A common noun is any noun *except* a proper noun. Said another way, common nouns name a general type of person (doctor, librarian), place (country, desert), thing (chemical, building), and idea (beauty, bravery).

We can now suggest two useful rules. First, proper nouns are virtually always capitalized, whereas common nouns are not. Second, proper nouns, being specific, are usually singular; common nouns can be either singular or plural. There is only one Mississippi River, but there can be one river or many rivers. There are many lakes, but only one Lake Michigan; however, Lake Michigan is one of the Great Lakes. The "Great Lakes" is plural in construction, but there is only *one* set of Great Lakes in the world.

In scientific writing, it helps in many ways to keep in mind the distinction between proper and common nouns. Two frequent problems faced by scientists are the distinctions between generic names and proprietary names of manufactured products and pharmaceutical preparations, and between scientific names and vernacular names of organisms. Here are some examples, with the proper nouns on the left, common nouns on the right.

| | |
|---|---|
| Xerox machine | photocopier |
| Doxycycline | tetracycline |
| Dacron | synthetic polyester |
| *Streptococcus* | streptococci |
| *Neisseria gonorrheae* | gonococcus |

Note that final example. *Neisseria* is capitalized, because there is only one genus *Neisseria*. The species name *gonorrheae* is not capitalized, perhaps on the grounds that there are a number of species within the genus *Neisseria*. These "proper-common" distinctions do not always hold up, but they usually provide a good guideline to capitalization.

## Concrete and Abstract Nouns

It is sometimes useful to keep in mind that common nouns can be subdivided again, into "concrete nouns" and "abstract nouns." The concrete nouns are those persons, places, or things that we can detect with our five senses (e.g., university, apple). Abstract nouns are those nouns, usually ideas or concepts, not directly detected by our senses (e.g., peace, friendship). Normally, these types of nouns are not troublesome.

## Collective and Mass Nouns

Two special types of common nouns are troublesome: collective nouns and mass nouns.

A collective noun indicates a group or collection of persons, places, things, or qualities (audience, committee, personnel, army, class). The general rule is that such nouns are plural in meaning but singular in form:

> The audience is restless.

> The couple owns a condominium.

Unfortunately, this rule often breaks down. Whenever the individuality of members of a group is emphasized, the plural form of the verb is used.

> The couple do not live together.

> The committee were from several scientific disciplines.

The best rule for handling collective nouns is to decide whether the *meaning* is singular or plural. Which of the following two sentences is correct?

> A total of 48 petri dishes were in the autoclave.

> A total of 48 petri dishes was in the autoclave.

Scientists who have a poor knowledge of English grammar would choose the verb "were," thinking that the subject of the sentence is "dishes." Scientists with a good knowledge of English would choose "was," recognizing that "dishes" is the object of a preposition and that the subject of the sentence is the singular word "total." Scientists with an *excellent* command of English would apply the *rule of meaning* and would select "were." To determine "meaning," we must ask ourselves what was in the autoclave. Was it the singular "total" or was it a whole mess of petri dishes? Obviously, it was the dishes; thus, the verb "were" is correct.

Another collective noun that comes up frequently in scientific writing is "number." Do we say "A number of test tubes is on the table"? No. Following the rule of meaning and recognizing that the plural word "tubes" is proof that more than one test tube is on the table, we say "A number of test tubes are on the table."

But, while following my simple rule of "meaning," do not simplistically conclude that words like "total" and "number" always take plural verbs. Look at these sentences:

A number of test tubes is on the table.

The number of test tubes on the table is four.

The first example, as already stated, is wrong; the verb "are" is needed to give logical meaning. However, the word "is" in the second example is correct. Why? Because there is only one number "four." Actually, the distinction here is caused by the difference between the definite article "the" and the indefinite article "a" (see Chapter 8).

Another confusing type of noun is the "mass noun." A mass noun is a concrete noun that represents a mass rather than countable units. Mass nouns are singular; many do not have plurals (air, water, wheat).

One of the most common grammatical errors is the misuse of the mass noun "amount" in place of the word "number." "An amount of people" is poor English, because people are countable individuals. We should say "The number of people on the elevator is nine." (Conceivably, we could weigh the people on the elevator. Then it would be correct to say "The amount of people on the elevator was 1,400 pounds.")

A related problem is the choice between "fewer" and "less." We would use "less" to modify nouns that can't be counted, and we would use "fewer" to modify a noun with countable units.

This beer has less taste.

This beer has fewer calories.

We cannot have "less calories" because, as scientists know, calories are countable units. The fact that commercials for a well-known beer are constantly yammering about "has more taste, has less calories" only shows that the TV commercials are tasteless as well as ungrammatical.

## Functions of Nouns

In sentences, nouns usually do something or something is done to them. A noun that does something is the *subject* of the sentence. If something is done to the noun, it is the *object* of a verb or of a preposition. (A preposition is a word used to relate a noun or a pronoun to some other part of the sentence.)

John hit the ball.

The proper noun *John* is the subject of the sentence; *ball* is the object of the verb *hit.*

> John hit the nail on the head.

Again, *John* is the subject, *nail* is the object of the verb, and *head* is the object of the preposition *on.*

In some sentences, nouns don't *do* anything nor is anything done to them. Such sentences usually present definitions or characteristics of these nouns. Typically, these sentences contain some form of the linking verb *to be.*

> Penicillin is an antibiotic.
>
> Scientists are nice people.

## PRONOUNS

A *pronoun* is a word used to replace a noun. The noun that the pronoun replaces is called the *antecedent.* Pronouns can be a bit tricky, because there are six different types of pronouns (personal, demonstrative, relative, interrogative, indefinite, and reflexive) and they have different forms that are easy to confuse.

### Personal Pronouns

A personal pronoun replaces a "person" noun. The form of the pronoun changes, depending on whether the pronoun is used as a subject, an object, or a possessive. The personal pronouns are:

> I, me, my, mine
> you, your, yours
> he, him, his

she, her, hers
it, its
we, us, our, ours
they, them, their, theirs

The personal pronouns cause comparatively few problems; we learn about "his" and "hers" during toilet training. When writing or revising, however, you are wise to examine each pronoun to make sure that it has an appropriate antecedent. Otherwise, you might write such confusing sentences as this one about (seemingly) human kidneys in dogs:

> No one yet had demonstrated the structure of the human kidneys, Vesalius having examined *them* only in dogs.

There is a problem with use of "sexist" pronouns (see Chapter 19), and *big* problems with *it, its,* and *it's.* The word *it's* is a contraction of "it is," and this contraction is often mixed up with the possessive pronoun *its.*

> Wrong: It's fur is fuzzy.
>
> Right: Its fur is fuzzy.
>
> Wrong: Its not good science.
>
> Right: It's not good science.

In addition to being "wrong" grammatically, you also risk misinforming your reader if you choose the wrong *its,* as in this pair of examples:

> A dog knows *its* master.
>
> A dog knows *it's* master.

The big problem with *it* is that the antecedent may be unclear. Unlike the other personal pronouns, the neutered "it" can stand in for virtually any noun in the sentence. So, watch "it" or you may be guilty of writing sentences like this:

> It is all right to give raw milk to your baby, but first boil it.

## Demonstrative Pronouns

A demonstrative pronoun singles out the thing referred to. These are ubiquitous in English writing. (Note that "These" in the preceding sentence is a demonstrative pronoun, the antecedent being the heading "Demonstrative Pronouns.") There are only four demonstrative pronouns: this, that, these, and those.

> *This* is my day.
>
> *That* is a crock.

*These* won't do.

*Those* are O.K.

These four words are not only used thousands of times as demonstrative pronouns, but they are also often used as adjectives (as in "these four words" in the first clause of this sentence).

## Relative Pronouns

Relative pronouns substitute for nouns *and* connect parts of sentences. These are the common relative pronouns: who, whom, which, whose, that, what, whatever, whoever, whomever.

> The laboratory director, *whose* office was on the second floor, was responsible for all research activities.

Note that the word "that" can be either a demonstrative pronoun (see above) or a relative pronoun. As a relative pronoun, *that* is often confused with *who*. Properly, *who* should be used to replace people, and *that* should be used to replace animals or inanimate objects.

> The man *who* came to dinner did not eat broccoli.

> I saw the cat *that* chased the rat.

> Conscience is the inner voice *that* warns us somebody may be looking.
> —H. L. Mencken

## Interrogative Pronouns

Interrogative pronouns are essentially the same as relative pronouns, except that the interrogative pronouns ask questions. The common ones are who, whom, which, whose, and what. Their purpose is to introduce questions.

> Who goes there?

> What happened?

## Indefinite Pronouns

Some pronouns are "indefinite," in that they replace nouns but not a *particular* person, place, or thing. Examples are the following: all, another, any, anyone, anything, both, each, either, everybody, few, many, most, much, neither, nobody, none, several, some, and such.

> Anyone can be lucky, but few succeed.

## Reflexive Pronouns

The least-used type of pronoun is the reflexive pronoun (myself, yourself, herself, themselves, etc.). However, these pronouns are frequently misused. Correctly used, a reflexive pronoun reflects the action of a verb back on the subject.

> I hit *myself.*

It is incorrect to use a reflexive pronoun either as a subject or as the object of a preposition.

> Wrong: John and *myself* will go home.
>
> Right: John and *I* will go home.
>
> Wrong: He hit John and *myself.*
>
> Right: He hit John and *me.*

Reflexive pronouns are sometimes used as "intensives," words used to intensify meaning or resolve.

> I *myself* will do it.
>
> I'd rather do it *myself.*

---

The rules are simple: The "-self" or "-selves" words are used for two purposes: first, to emphasize ("Mother, I'd rather do it myself"), and second, reflexively, so that the action is turned back on the grammatical subject:

I never quite accustomed myself to the altitude of Denver.

The dermatologist learned that the patients had overexposed themselves to ultraviolet radiation from tanning devices.

—Edith Schwager

---

# Chapter 7
# Action Words (Verbs)

*For most purposes the best writing is direct writing—writing that avoids three words where two will do, writing that represents an action in a verb and the agent of that action as its subject.*
— Wilma R. Ebbitt and David R. Ebbitt

## FUNCTION OF VERBS

In the preceding chapter, I discussed nouns and pronouns. As the subject of a sentence, something that has been named (a noun or a pronoun) is likely to be followed by a word that does one of two things: describes the existence or a characteristic of the named subject, or describes an action of the subject.

> She *is* intelligent.
>
> He *hit* the ball.
>
> The chemist *added* hydrochloric acid.

## TYPES OF VERBS

The "existence or characteristic" verb is often some form of "to be" (is, are, was, were, etc.), but other "linking" verbs (e.g., become, taste, smell, grow) are used frequently.

> Sulfuric acid *is* a common reagent.
>
> I *am* lonely.
>
> This *tastes* terrible.

All other verbs can be divided into two types: transitive and intransitive. All this means is that some verbs (transitive) take an object (a noun following a verb); others (intransitive) do not.

> He *gave* me a hammer.
>
> She *ran*.
>
> They *died*.

In the first example, *gave* took an object (*hammer,* with *me* being an indirect object). The second and third examples have no objects; hence, the verbs *ran* and *died* are intransitive.

Some verbs can be used either transitively or intransitively.

> He *grows* roses.
>
> The guinea pigs *grow* well.

In the first example, *roses* is a noun, the object of the transitive verb *grows.* In the second example, the only word after the verb is the word *well,* which is an adverb, not a noun; thus, the sentence has no object and *grow* is intransitive.

## NOMINALIZATIONS

One of the most frequent faults of scientists as writers is that they often confuse actions (verbs) with agents of the action (usually nouns). Such failures result in sentences that are difficult at best and incomprehensible at worst. As I see it, the three most common types of agent-action confusion result from (1) the reluctance of scientists to use first-person pronouns, (2) the overuse of the passive voice, and (3) the regrettable tendency to turn sharp action words (verbs) into weighty nouns. The first two of these faults are dealt with in more detail in Chapter 15. The third, called "nominalizations," is castigated here. (Nominalizations are castigated again in Chapter 10. Such a cardinal sin merits redundant sermons.)

If I say "I studied the effect of A on B," you know what was done, you know who did it, and you know approximately when it was done (probably recently, certainly during the lifetime of "I"). However, if we turn the verb "studied" into a noun ("study" or, in the unfortunate jargon of scientists, "investigation"), if we omit the agent of the action ("I"), and if we put the whole thing into the passive voice, we have something like this:

> An investigation was undertaken to determine the possible effect of A on B.

Now, in more words, we have a sentence of a type that is all too typical in science, a sentence that says almost nothing. We no longer know that A affects B; we know only that there is a "possible effect." We do not have the foggiest idea of who "conducted the investigation" (another awful phrase scientists love to use); it might be I, it might be my Aunt Min, or it might have been Julius Caesar. Nor do we know when this "observation was made" (another awful phrase; why not use the active voice "I observed" or "they observed"?); the "observation" might have been "made" last week, 20 years ago, or 2,000 years ago.

"We rejected that theory." Now that is a clear statement. Turn the verb into a nominalization, and we have "The rejection of that theory has been reported."

Get into the habit of looking for nominalizations; many end with "tion." Almost always, the action verbs will give the clearest meaning. We *investigated* is much better than an anonymous *investigation.* They *produced* is better than a vague *production. Consumed* is usually better than *consumption.* I *informed* him is good. I *told* him is better. The *information was communicated* to him is terrible. Look carefully at the next two examples.

> The installation of the new computer can be performed in 3 days.

> We can install the computer in 3 days.

You should agree that the second example is much better than the first.

---

Language is in decline. Not only has eloquence departed but simple, direct speech as well, though pomposity and banality have not.

—Edwin Newman

# Chapter 8
# Descriptive Words
# (Adjectives, Adverbs,
# and Articles)

*I'm glad you like adverbs—I adore them; they are the only qualifications
I really much respect.*

—Henry James

## THE DESCRIBERS

The agents of the action (nouns, pronouns) can be described or qualified.
These modifiers are called *adjectives*. The action (verb) can also be
modified. These modifiers are called *adverbs*. Even modifiers can be
modified (by adverbs). In English and some other languages (but not
Japanese, for example), we have a group of words called *articles*. In English,
these are the three words *a, an,* and *the.* They are often treated as adjectives,
but their usage is so peculiar that they are treated separately here.

## ADJECTIVES

An adjective modifies a noun or a pronoun.

A *red* apple.

The apple is *red*.

Usually, adjectives precede the nouns they modify (as in the first
example). Sometimes, adjectives follow the noun and a linking verb (as in
the second example).

Adjectives not only state a quality of the noun or pronoun they modify, but they can also be used with various degrees of intensity. Each adjective can be "compared" as follows:

cold (positive)

colder (comparative)

coldest (superlative)

Unfortunately, some English adjectives are "irregular" and do not follow the "er" "est" style:

much (positive)

more (comparative)

most (superlative)

Words can also be "compared" without the use of the "er" and "est" endings, even if the adjectives are not irregular. Instead of cold, colder, coldest, we can say cold, more cold, and most cold.

Finally, English has a number of "absolute" words that are not subject to comparison, e.g., unique, perfect, exact, and infinite. Something is either "unique" or it isn't; it can't be more or less unique than something else.

## ADVERBS

Adverbs are words that modify verbs, adjectives, or other adverbs.

He went *slowly*.

It was *very* small.

He went *very* slowly.

In the first example, the adverb modifies a verb; in the second, the adverb modifies an adjective; in the third, the adverb "very" modifies the adverb "slowly."

Many adverbs end in "ly" and are thus easy to identify. Others do not and thus are not. Many adverbs have related adjectives with which they can be confused. For example, the adjective "real" is often used inappropriately in place of the adverb "really":

This soup is *really* good.

This soup is *real* good.

The first example is correct; the second is substandard usage; either *really* good or *very* good is preferable.

I feel *bad*.

I feel *badly*.

Both of these examples are correct (if used properly), but they mean different things. In the first example, *bad* is an adjective (modifying *I*), and the sentence means that I feel lousy. In the second example, *badly* is an adverb (modifying *feel*), and the sentence means that I have poor tactile sense.

With all modifiers, the key to good usage is to get them close to the words they modify. Adjectives usually fit into place easily, because the nouns or pronouns they modify are usually clear. Adverbs, however, can modify verbs, adjectives, and other adverbs. Thus, their exact placement in a sentence is important to the meaning of the sentence. Adverbs such as *only, often,* and *never* are often misplaced in scientific writing, leaving sentences that are unclear. (*See* Chapter 4.) Precision with English requires careful word order, which is known as syntax. In every sentence, the agent of the action must be clear, the action or state of being must be clear, and any modifiers used to describe the agent of the action or the action (or the object of the action) must be carefully placed.

## ARTICLES

The articles *a, an,* and *the* are the most common words in scientific or any other kind of writing. The words *a* and *an* are called "indefinite articles"; *the* is called the "definite article." The indefinite article *a* is used before words that begin with a consonant sound. The article *an* is used before words starting with a vowel sound.

*a* noun

*a* woman

*an* apple

*an* orthopedic procedure

It doesn't matter whether the following word is a vowel or a consonant; what matters is the *sound.* This rule is particularly needed in placing an article before an abbreviation.

*a* Master of Science degree

*an* M.S. degree

Both of these examples are correct. We say *a Master* because the "M" in "Master" is obviously the consonant sound, but we say *an M.S.* because we pronounce the *M.S.* as "em ess," meaning that the *sound* is now the vowel *e*. Likewise, *an mRNA* is correct because the reader will read "emRNA," not "*messenger* ribonucleic acid."

The British tend to put *an* before such words as "historian," whereas most Americans would write "a history" rather than "an history." On the other hand, in a word such as "honor," the "h" is silent and "an honor" is the correct usage.

Articles, if used carefully, can serve as guides to the reader. The primary purpose of an article is to identify a noun. Because a huge number of words in English can be either nouns or verbs, we need the articles to point to the nouns. Without an article, this sentence is confusing.

Plan moves slowly.

In this sentence, two of the three words (plan, move) are words that can be either nouns or verbs. As written, this sentence probably means only "[You should] plan moves slowly." But note how the addition of an article clarifies the meaning:

Plan the moves slowly.

The plan moves slowly.

The distinction between the indefinite articles and the definite article can sometimes provide helpful diagnosis. Recall that collective nouns can take either singular or plural verbs. However, the choice of article can make clear which verb form is needed.

*A* number of apples *are* on the table.

*The* number of apples on the table *is* 14.

In the first example, "a number" is almost certainly meant as more than one; thus, it sounds plural, and in this case it is. On the other hand, "the number" signifies one number (even though it may be a large number); thus, it sounds singular, and in this example it is.

---

The adverb *very* has become emasculated through overuse. I would make this suggestion: Every time you have the impulse to write *very*, repress the impulse and see whether the omission would entail any real loss of meaning. At first you will almost certainly say, Yes, something is lost. But I predict that if you persevere, you will gradually agree that any loss is negligible. *Very* is one of the words that contributes to flabby writing.

—Lester S. King

# Chapter 9
# Function Words
# (Conjunctions,
# Prepositions, and
# Interjections)

*A preposition is a poor word to end a sentence with.*

—Anonymous

## JOINING THE PIECES

English has agents of action (nouns and pronouns), action (verbs), and descriptions of the agents (adjectives) and the actions (adverbs). In addition, English has some words which have little meaning but which perform functions. Such words often act as a glue holding parts of a sentence together.

## CONJUNCTIONS

Conjunctions are used to connect words, phrases, or clauses.

Joe *and* Mary are going to the party.

She is not a friend of mine *nor* of his.

He ran, *but* she walked.

Some conjunctions express equal weights of the joined words, phrases, or clauses; these are called *coordinating conjunctions*. There are seven coordinating conjunctions: and, but, or, for, nor, so, yet. These words connect the two clauses of a compound sentence: He ran, *but* she walked. Because these seven words are among the most commonly used English words, and because your ability to punctuate sentences depends upon your ability to recognize these seven coordinating conjunctions, I recommend that you memorize them. As a mnemonic device, you might try remembering the acronym FANBOYS, which stands for *for, and, nor, but, or, yet,* and *so.*

Another class of conjunctions is called *subordinating conjunctions.* These connect *unequal* parts; for example, they are used to connect an independent clause with a dependent clause. (Don't get panicky. Clauses will be fully explained in Chapter 12.)

> Joe went to a party *after* he left the office.

A clause introduced by a subordinating conjunction is a *subordinate* (or *dependent)* clause. These conjunctions often indicate a time relationship or some other limiting function. The most common are the following: although, before, after, because, if, where, than, since, as, unless, that, though, when, whereas, while.

> *When* the party was over, he went home.

Still another joiner is the *coordinating adverb.* These are like the coordinating conjunctions in that they are used to connect independent clauses. However, coordinating conjunctions are preceded by a comma, whereas coordinating adverbs are preceded by a semicolon and followed by a comma. Coordinating adverbs include the following: however, moreover, therefore, further, then, consequently, besides, accordingly, also, too.

> Joe decided to go to the party; *however,* Mary decided to go home.

> We ran out of beer; *therefore,* the party was over.

But do not get carried away and put a semicolon before every *however,* because *however* and other coordinating adverbs can also serve as normal adverbs and are punctuated accordingly.

> I hope you learn this lesson, *however* long it takes.

> The yield is *therefore* small.

## PREPOSITIONS

Prepositions combine with nouns or pronouns to form a phrase. There are about 70 prepositions in the English language, most of them expressing direction or location.

> *to* the right

> *in* the middle

Scientists and others have four main types of problems with prepositional phrases. The first, as with other building blocks of English, is syntax (word order). If the phrase is not close to the noun or pronoun it modifies, we may have syntactic mayhem, as in the classic:

> For sale, car owned by lady with dent in rear.

The above short sentence fragment has four prepositional phrases, showing how very common these are in English. The first ("For sale") is O.K., because it is immediately followed by "car," the item for sale. The "by lady" follows "owned" and the "in rear" follows "dent," as they should. However, the "with dent" is much too far from "car," which it presumably modifies, and much too close to "lady," which it does not modify (unless the lady is indeed unfortunate).

A second but frequent problem is the doubling of prepositions.

> *Inside of* the park, many animals lived.

> That question is *outside of* my field of expertise.

Because "inside" and "outside" are prepositions, as is "of," the doubling of prepositions provides redundancy. The phrases should read "Inside the park" and "outside my field." Here is a similar example, from TV's "Murder, She Wrote": "Arnold raced *out of* the door." If Arnold were a termite, perhaps he could have "raced out of the door."

The third problem often afflicting prepositions has to do with *case.* Prepositions are usually followed by *objects.* Thus, prepositions always take the objective case, never the nominative. *(Objective* relates to objects; *nominative* relates to subjects.)

> The argument was between he and I.

In this example, the preposition *between* is erroneously followed by the nominatives *he* and *I.* The preposition should be followed by pronouns in the objective case, and the sentence should read: "The argument was between *him* and *me.*"

The fourth common problem is the use of a preposition to link an adjective (rather than a noun or pronoun) to another part of a sentence. The *of* should be deleted from the following sentence.

> She is too good *of* a person to complain.

Some grammarians have argued that a preposition should not end a sentence. It was this rule that prompted Winston Churchill's famous rejoinder: "This is the kind of nonsense up with which I will not put." Not to worry. Modern grammarians agree with Churchill. As for me, I see nothing terribly wrong with the following sentence, which ends with *four* prepositions. A child asks: "Why have you brought me that book to be read to out of for?"

On the other hand, sentences ending with a preposition are often awkward.

> Chemical engineering was what he took his degree in.

Such sentences can easily be recast to make shorter, clearer sentences.

> He took his degree in chemical engineering.

And, of course, you have heard about the prostitute who made the mistake of propositioning a plainclothes policeman. Her proposition ended with a sentence.

## INTERJECTIONS

An interjection is a word, phrase, or sentence expressing emotion.

> Hey!

> Of course!

> I thought so!

Strong interjections (those followed by exclamation points) are rarely used in scientific writing. Mild interjections, which are usually separated from the rest of the sentence by commas, are used occasionally.

> Oh, *well*, it was worth a try.

> Waksman, *indeed*, was the discoverer of streptomycin.

If I might interject a thought about interjections, it would be: Don't use them!

---

Interjections are needed in conversation to fill gaps—some of us say "uhh" and some use interjections, to about the same effect—but writers do not leave gaps, assuming they take the time to think through their ideas and to rewrite for succinctness.

—Gregory A. Barnes

# Chapter 10
# Prefixes and Suffixes

*I find the growth in use of the prefix "pre" in biology especially galling. It seems to have spread, like a plague, from manufacturers' catalogs to methods sections. It is used in conjunction with many words; some of the most common are pretreated, prewashed, presoaked, and presterilized, which mean, of course, treated, washed, soaked, and sterilized.*

—Peter Kulakosky

## PEAT AND REPEAT

In the preceding chapters, I have briefly described the various kinds of words in the English language. English, being the incredibly rich language that it is, further adds to its vocabulary a large number of bits and pieces of words; we call these prefixes and suffixes. We can use these prefixes and suffixes to give different meanings to words. Thus, we can take the form of a particular word that best suits our meaning. Take that word *take.*

We can *take* it once.

If we can take it once, it is *takeable.*

If we do it again, we can *retake* it.

Having shown that we can do it again, we know that it is *retakeable.*

If we can't do it at all, it is *untakeable.*

Unfortunately, these beginnings (prefixes) and endings (suffixes) do not fit every word. When a word starts with *re*, we cannot assume that a prefix is being used (as indicated by the heading of this section, "peat and repeat"). The *re* in *repeat* is not a prefix; nor can we add *re* to *repeat*, because *rerepeat* would not make sense. Similarly, the *re* in *recover* is not a prefix when we *recover* a lost article; *re* is a prefix, however, when we *re-cover* our sofa. Note that this *re-cover* takes a hyphen to distinguish it from the other *recover*. (I hope that you will recover from all of this.) Probably all prefixes have similar exceptions. I can be *overstimulated* or just *stimulated*; I can also be *overjoyed* but I cannot be *joyed*.

The prefix *pre* can be used before a great many words (*pre*birth, *pre*examine, *pre*history) and even numbers (*pre*-1900). The very word *prefix* has *pre* as a prefix. But, of course, *pre* is not always a prefix. A doctor who uses good English could *scribe* for a patient, but most likely the message would be different if he were to *prescribe* for the patient.

## NEGATIVE PREFIXES

Perhaps the most useful use (to coin a phrase) of prefixes is to provide meanings exactly the opposite of the original meanings. A *political* person is intensely interested in politics; an *apolitical* person has no interest in politics. (Because *a* can be either a prefix or an article, oddities occur; *a theist* believes in God, whereas an *atheist* does not.) A *starter* can win a horserace; a *nonstarter*, never. It is good when I am *well;* it is not good when I am *unwell.*

## SPELLING

Prefixes and suffixes are normally set solid with no space or hyphen between the prefix or suffix and the root word. In a Mexican restaurant, the *refried* beans are more accurate if not more tasty than *re fried* beans or *re-fried* beans. The tendency in English seems to be to set prefixes solid even when a letter is doubled. A few grammarians still prefer hyphenation in such instances (*re-enter, co-operate*) or diaeresis marks (*coöperate*).

Therefore, get into the habit of joining the parts; you will then be right most of the time. Unfortunately, such joining on occasion creates entirely different words. For example, if you have a material that was *un-ionized*, we know what you mean; if it was *unionized*, don't forget to sew on the union label.

Remember that these rules apply to real prefixes (such as *re, co,* and *pre*). However, some "prefixes" are complete words that often precede others. Take *sea* as an example. Some *sea* word combinations are separate from each other (sea bass, sea breeze, sea level); other combinations are joined (seabird, seacoast, seaport). How can you tell which are joined and which are not? The sure way is to check the dictionary. However, such word combinations come up so often that you won't get much writing done if you stop to check each combination. You can make the right determination almost every time *if* you know how the combinations are pronounced. If the two parts are stressed equally, they are separated by a space (sea lane, sea power). If the first word is stressed, that "word" has become an accented syllable and the two "words" have become one (seafood, seashell).

## CONFUSING PREFIXES

Some prefixes are confusing. The word *flammable* probably refers to something that can catch on fire; theoretically, something *inflammable* would not catch fire. However, our mind's eye remembers that the word *inflame* means to set fire, and we thus worry about a substance that might be "highly inflammable." When such confusion arises, it is best to dispense with the prefixes: say *flammable* or *not flammable*.

## SUFFIXES

Frequently, suffixes are added to verbs to turn them into nouns. The verb *develop* becomes *development*; *govern* becomes *governance*; *reject* becomes *rejection*; and so on. These nouns are often useful. However, a strong and regrettable tendency in scientific writing is to use the noun form, covering up the real action going on in a sentence. These nouns (called nominalizations; *see also* Chapter 7) are often combined with the passive voice, resulting in wordy sentences in which both the agent of the action and the action itself are hidden in the underbrush:

> The rejection of the manuscript was performed by the Editor.
>
> Our results were in agreement with theirs.

By adding action to these two sentences (and reducing the number of words) we have:

> The Editor rejected the manuscript.
>
> Our results agreed with theirs.

Contrariwise, nouns can be turned into verbs by adding a suffix:

idol          idolize

type          typify

Such words can be useful. However, some suffixes (*ize*, for example) are used to create jargon (blenderize). The use of *wise* is often appropriate (*lengthwise*); sometimes it is questionable (*policywise*); sometimes it is silly jargon (*philosophywise*).

I will end this chapter with a bit of nonsense (prefix *non* added to *sense*):

Adults have more fun in adultery than infants have in infancy.

# Chapter 11
# Phrases

*There is nothing wrong with splitting an infinitive ("He is going to about make the grade") except that eighteenth- and nineteenth-century grammarians, for one reason or another, frowned on it. And most grammar teachers have been frowning ever since.*

—Theodore M. Bernstein

## KINDS OF PHRASES

The words from the nine pigeonholes (nouns, pronouns, verbs, adjectives, adverbs, prepositions, conjunctions, interjections, articles) of the English language can be assembled into phrases, clauses, and, ultimately, sentences, the basic units of thought and communication. Ah! But how do we string these words together?

First, some definitions. A phrase is a group of two or more grammatically related words that do not make a full statement, i.e., a group of related words that does not include both a subject and a verb. On the other hand, a clause (*see* Chapter 12) does contain both a subject and a verb.

If a phrase is a group of related words, we could say (and grammarians do) that a noun with its associated adjectives is a "noun phrase." Thus, we can have noun phrases, verb phrases, adverbial phrases, and a bewildering taxonomic array. It is better, I think, to let adjectives quietly modify their nouns, for example, without worrying about noun phrases. Instead, let us look at those groups of words that, *as a group,* act as a part of a sentence or modify some other element in a sentence. Fortunately, there are only four of these: prepositional phrases, infinitive phrases, participial phrases, and gerund phrases. In assembling these phrases into sentences, there is one overriding rule: Get them as close as possible to the sentence elements they modify.

# PREPOSITIONAL PHRASES

A prepositional phrase consists of a preposition, its object, and any words that modify the object. Normally, prepositional phrases act as adverbs or adjectives

*For dinner*, he went *to the most popular and most expensive restaurant in the city.*

In the above example, only the two words "he went" are not parts of prepositional phrases. The sentence has three prepositional phrases: *For dinner* modifies "went" and is thus serving as an adverb; *to the most popular and most expensive restaurant* is a much longer prepositional phrase, also modifying "went"; *in the city* is a prepositional phrase modifying the object of the preceding prepositional phrase ("restaurant"), and it is thus serving as an adjective.

Note that "For dinner" in the above example is followed by a comma. Here is a rule: When a sentence starts with a preposition, insert a comma after the prepositional phrase. Many grammarians would argue that such commas are not necessary, particularly if the phrases are short. However, the use of such commas *is never wrong,* and use of commas after every introductory prepositional phrase will provide consistently clear sentences. In the example above, no confusion results if a comma is not used following "For dinner." However, try this one:

For the first time scientists have been able to treat this disease.

If no comma is used after the opening prepositional phrase ("For the first time"), the reader for a moment may mistakenly assume that the sentence refers to "first time scientists." Thus, even though introductory prepositional phrases need not always be followed by a comma, insert the commas anyway. Then you need not worry about when to use a comma and when not, and you will never be wrong. In my undivided opinion, this is a good rule.

# INFINITIVE PHRASES

An infinitive phrase consists of "to" plus a verb and any objects or modifiers.

*To pass this course,* you must be willing *to study diligently.*

Infinitive phrases are easy to use. The only rule: Place the infinitive phrase as close as possible to the word it modifies. And *forget* the old rule against splitting infinitives. (A "split infinitive" has one or more words between the "to" and the verb.) That is not to say that infinitives should always or even usually be split. In the above example, for example, "to study diligently" perhaps flows more smoothly than "to diligently study."

The treatment failed to delay further signs of infection.

In the above example, however, the reader may infer that "delay" will lead to "further" (additional) signs of infection. If we split the infinitive (to further delay), the reader knows we mean additional delay and not additional signs of infection. I have seen many thousands of sentences that were confusing because their authors were afraid to split infinitives.

As for me, I go out of my way to carefully split them every day, and I am proud of it. (Split infinitives are treated in greater detail in Chapter 4.)

## PARTICIPIAL PHRASES

A participial phrase consists of a participle (an "ing" verb), its object, and its modifiers.

Using good sense, you can write good sentences.

A participial phrase functions as an adjective, so it must modify a noun or a pronoun; in addition it must be close to the word it modifies. If the sentence contains no noun or pronoun that the participle can modify, or if the noun or pronoun is too far away from the participle, the participle is said to "dangle."

My favorite dangling participle is this one:

Lying on top of the intestine, you can perhaps make out a thin transparent thread.

Because a participle *should* modify the nearest available noun or pronoun, the unwary reader of the above sentence would assume that "you" are "Lying on top of the intestine." Presumably the writer was trying to say that the "thread" was on top of the intestine. As written, the very first word in this sentence (the participle "Lying") actually modifies the very last word in the sentence (the noun "thread"); thus, syntactically, the sentence could not be more wrong. The problem is solved by turning the sentence around:

You can perhaps make out a thin transparent thread lying on top of the intestine.

Most dangling participles occur when the main clause following the participle begins with "it is."

Going home, it is wise to be careful.

© 1992, Washington Post Writers Group. Reprinted with permission.

## GERUND PHRASES

A gerund is an "ing" verb used as a noun. A gerund can be used alone.

Running is my favorite sport.

Or, by giving the gerund an object (with or without modifiers), we can create a gerund phrase. Such gerund phrases can be used as the subjects of sentences.

Writing this chapter was a difficult task.

Gerund phrases can also be used as objects.

I enjoyed writing this chapter.

Fortunately, gerund phrases seem to fall naturally into place when used in sentences. Rarely do they cause the kinds of confusion so often caused by participial phrases.

> Every single phrase is a string of perfect gems, of purest ray serene, strung together on a loose golden thread.
>
> —George du Maurier

# Chapter 12
# Clauses

*Sentences beginning with* and, but, *or* or *are acceptable provided the practice is not overdone. Indeed, nothing that is overdone is ever acceptable, be it language or lamb chops.*

—Theodore M. Bernstein

## KINDS OF CLAUSES

A clause functions as a part of a sentence; unlike a phrase, however, a clause has both a subject and a verb. Fortunately, clauses are easy to understand because there are only two main types of clauses: independent and dependent.

## INDEPENDENT CLAUSES

An independent clause is a set of words, including a subject and a verb, that makes a complete statement. Thus, an independent clause could stand alone as a complete sentence. However, if it stood alone, it would be called a sentence and not a clause. As a clause, it is attached to some other sentence element (another independent clause or a dependent clause).

No amount of experimentation can prove me right; a simple experiment may at any time prove me wrong.          —Albert Einstein

The above sentence includes two independent clauses separated by a semicolon. Note that the semicolon could be replaced by a period (full stop), and the two independent clauses could indeed stand as sentences. Einstein, however, wanted to show a close relationship—a contrast—between these two related thoughts; therefore, he expressed these thoughts as two clauses rather than as two sentences.

Independent clauses are usually separated from each other in one of three ways. First, a semicolon can be placed between the two clauses.

Life has not just passed me by; it gave me a kick in the pants while passing.

Second, a semicolon plus a coordinating adverb (e.g., however, therefore, moreover) can be used; when independent clauses are joined in this way, a semicolon always precedes the coordinating adverb and a comma always follows it.

I tried to find my way out of the forest; however, I became hopelessly lost.

Third, two independent clauses can be joined by a comma and a coordinating conjunction (and, but, for, nor, or, so, yet).

I went to the store, and I bought a loaf of bread.

Two independent clauses can *not* be joined by a comma without a coordinating conjunction.

Seek wealth, it's good.
—Ivan Boesky

The error in the above sentence is called a "comma splice." (Mr. Boesky spent three years in prison, but I have forgotten whether his sentence was for his grammatical error or for following his own advice.)

## DEPENDENT CLAUSES

A dependent clause, like an independent clause, has both a subject and a verb. However, a dependent clause is introduced by a subordinating word which makes the clause dependent on (subordinate to) another (independent) clause. Many of these subordinating words express time constraints (when, after, before) or doubt (if, whether).

If the Republicans will stop telling lies about the Democrats, we will stop telling the truth about them.                                    —Adlai Stevenson

If it weren't for my plumber, I wouldn't have any place to go.

When scientists attempt to predict earthquakes, they are on shaky ground.

When sex is good, it is very, very good.

In the final example, the independent clause is "it is very, very good"; the dependent clause is "When sex is good." Obviously, the dependent clause could not stand alone as a sentence. Note that a comma follows the dependent clause, indicating to the reader that the dependent clause is ending and that the independent clause is beginning. This is a *rule*: When a sentence begins with a dependent clause, end the clause with a comma. The rule can be extended to indicate that a comma should follow any dependent clause preceding an independent clause. A double-header example is given in the following sentence:

When sex is good, it is very, very good; when it is bad, it is still good.

Actually, the above *rule* is not always correct because (rarely) a dependent clause can be used as a subject of an independent clause; since it is a cardinal rule that a comma cannot separate a subject from its verb, a comma is not used.

That you are enjoying this book is very likely.

The above is a kind of reverse sentence. (Most people would probably write "It is very likely that you are enjoying this book.") Nonetheless, the example is grammatically correct; "That you are enjoying this book" is serving as the subject of the sentence. Note also that, in the parenthetical sentence, "that you are enjoying this book" is used as a predicate adjective. In fact, dependent clauses can serve as nouns (as in the first example), as adjectives (as in the revised example), and as adverbs:

We will all go together *when we go*.

In this example, note that a comma is not used before the dependent clause, whereas a comma would be used if the dependent clause started the sentence:

When we go, we will all go together.

When a dependent clause appears in the *middle* of an independent clause, two commas (balancing commas) are usually used, one before and one after the dependent clause.

A cynic is a man who, when he smells flowers, looks around for a coffin.
—H. L. Mencken

She, whom I have known for many years, arrived yesterday.

Note that the "whom" clause has commas fore and aft. Note also that the clause modifies the pronoun "she," which means that the clause is adjectival. (By definition, adjectives modify nouns or pronouns.) Finally, in passing, note that the objective case "whom" is used. It appears that "whom" is slowly passing out of the language; many writers today would say "She, who I have known for years, arrived yesterday." However, if one can distinguish between the nominative and objective cases, it still usually sounds better to use "whom" as an object. Admittedly, in many "cases," as in the above example, it is not easy to distinguish between the cases. In the above example, the normal subject-verb-object syntax (I have known whom) has been inverted (who or whom I have known), making the "who" look like a subject even though in reality it is an object.

## WHICH HUNTING

Many grammarians and editors love to play games with "which" and "that." Sooner or later, every author will have a manuscript returned by an editor who has changed all of the "whiches" in the manuscript to "thats" and vice versa. Such "which hunters" believe that dependent clauses can be divided into two types: restrictive and nonrestrictive (or defining and nondefining). By their definition, restrictive clauses (those that are essential to the meaning of a sentence) are preceded by "that"; nonrestrictive clauses (not essential) should be preceded by "which."

Unfortunately, as with other grammatical rules, the that-which rule is impossible to invoke across the board. Some dependent clauses are obviously restrictive, supplying essential information, and others are obviously supplying only incidental information. However, who can possibly define which modifiers are essential and which give only additive information? Normally, such distinctions should be made by *authors*, not *editors*, because only the author of a sentence knows whether something is "essential" to his or her message. And, often it simply doesn't matter, to the author or to anybody, whether "that" or "which" is used.

Beef is a meat (that, which) is high in protein.

This is an experiment (that, which) will succeed.

According to the "rule," "that" should be used in these sentences. (A sentence saying "Beef is a meat" wouldn't make much sense until the restrictive clause is added.) Does that mean "which" would be wrong in the above sentences? I have a better question: Who gives a damn? In zillions of such sentences, either "that" or "which" can be used and the meaning of the sentences will be exactly the same.

However, someone in the peanut gallery just raised a hand and said "Does that mean we can forget the that-which rule?" The answer is no. Absolutely not.

Antibiotics (that, which) are produced by microorganisms are effective in treating most infectious diseases.

Alkaloids (that, which) contain nitrogen can be poisonous.

Does it matter whether "that" or "which" is used in the above sentences? It certainly does. In the first sentence, if "that" follows "antibiotics," the implication is that some antibiotics are produced by microorganisms and others are not. If "which" follows "antibiotics" (and especially if a comma precedes "which" and a balancing comma follows "microorganisms"), the sentence then means that *all* antibiotics are produced by microorganisms. Likewise, in the second example, use of "that" would suggest that some alkaloids contain nitrogen and others do not, whereas use of "which" would indicate that *all* alkaloids contain nitrogen.

Therefore, you need not be a "which hunter," but you should occasionally watch for "whiches," especially on Halloween. In choosing between "which" and "that," you might try writing the sentence twice with a "which," once with commas around the clause, once without:

Alkaloids, which contain nitrogen, can be poisonous.

Alkaloids which contain nitrogen can be poisonous.

In this way, the distinction in meaning (if any) will become clear. If the material between the commas could be deleted from the sentence without destroying the sentence, the clause indeed is "nonrestrictive," and the "which" and the commas should be retained. The basic sentence "Alkaloids can be poisonous" now has a nonrestrictive addition: "Alkaloids, which contain nitrogen, can be poisonous." If, however, the material within commas cannot be deleted without ruining the sentence, the commas should *not* be used (and, ideally, the "which" should be changed to "that"). If your meaning is not that "alkaloids can be poisonous" but that only those alkaloids containing nitrogen can be poisonous, your meaning is then expressed clearly by saying "Alkaloids that contain nitrogen can be poisonous."

This means that good writers must know how and when to distinguish between "which" and "that."

---

It is almost always a greater pleasure to come across a semicolon than a period . . . . You get a pleasant little feeling of expectancy; there is more to come; read on; it will get clearer.

—George F. Will

# Chapter 13
# Sentences

*Vigorous writing is concise. A sentence should contain no unnecessary words, a paragraph no unnecessary sentences, for the same reason that a drawing should have no unnecessary lines and a machine no unnecessary parts. This requires not that the writer make all his sentences short, or that he avoid all detail and treat his subjects only in outline, but that every word tell.*

—William Strunk, Jr.

## KINDS OF SENTENCES

Words, phrases, and clauses are the building blocks of sentences. The sentence is the basic unit of thought and expression. If you can build sentences logically, you can communicate clearly. It is easy to build effective sentences because—in all of the millions of books, journals, and other written documents—exactly six types of sentences are used. Learn how to organize and punctuate these six types of sentences and you will be able to write clearly. These six types are questions, exclamations, and the four kinds of declarative sentences (simple, compound, complex, and compound-complex).

## QUESTIONS

A question is any sentence that asks for something and therefore ends with a question mark. True, the word "question" can be used more broadly (e.g., the question before the house), and some indirect questions are not followed by question marks. However, such "questions" become declarative in form, so it is best to simply accept the idea that questions end with question marks.

What is a question?

How should I know?

By further definition, a question is a sentence that needs a response.

How did I know the B-1 bomber was an airplane? I thought it was vitamins
for the troops.                                    —Ronald Reagan

What time is it?

If you ask somebody what time it is, you expect a response.

Would you turn in this assignment next Tuesday, please.

In the above example, you do not really expect an answer. In fact, you
are commanding a response, rather than asking for one. Although it *looks*
like a question, it is a declarative sentence (and thus it ends with a period).

## EXCLAMATIONS

An exclamation is any sentence which ends with an exclamation point.
While in the business of giving simple definitions, I will also define
declarative sentences (all four types) as those that end with periods. Now,
do you see how simple English is?

Exclamations can be single words, phrases, clauses, or full sentences.

Attention!

At the ready!

When I am good and ready!

Stick it in your ear!

Exclamation points lose their force when authors use them frequently.
They can also annoy readers when they are used to conclude seemingly
inconsequential remarks. In general, avoid using exclamations in scientific
writing.

## DECLARATIVE SENTENCES

### Simple Declarative Sentences

A simple sentence consists of one independent clause. That is, the
sentence contains a subject and a verb; it may also contain an object.

He ran.

She solved the problem.

We must keep America whole and safe and unspoiled.

—Al Capone

The normal structure of the simple declarative sentence is subject, verb, and object (if any) or predicate adjective (if any).

The world exists.

A meteorite hit the Earth.

She is intelligent.

Virtue is its own punishment.

—Aneurin Bevan

A simple declarative sentence is one unit of thought. Such straightforward, uncluttered statements should form the basis of good scientific writing. Other sentence structures are needed from time to time, but the expressive power of the simple declarative sentence is recognized by most good writers.

Simple sentences are not necessarily short. Modifying words and phrases can be added to subjects, verbs, and objects. Furthermore, simple sentences can have more than one subject (called a compound subject), more than one verb (a compound verb), and more than one object. All of the following are simple declarative sentences:

He ran.

He and she ran.

She ran and jumped.

He and she ran and jumped.

She painted the house, the barn, and the corncrib.

## Compound Declarative Sentences

Compound sentences contain two independent clauses. Almost any two simple sentences can be joined together to form a compound sentence. If the two thoughts in the two simple sentences are not closely related, the sentences should be retained as separate sentences; if there is a fairly close connection between the two thoughts, the two sentences should be fused into one.

The two independent clauses of a compound sentence are usually joined in one of three ways: (1) with a comma and a coordinating conjunction (and, or, but, for, nor, so, yet); (2) with a semicolon; or (3) with a semicolon and a coordinating adverb such as however, therefore, thus, moreover, or nevertheless.

The plumber located the problem, but she found it difficult to repair.

The plumber located the problem; she found it difficult to repair.

The plumber located the problem; however, she found it difficult to repair.

The above three punctuational choices are the standard ways of joining two independent clauses. Rarely, the second clause may relate very closely to the first clause, and the writer might want to show this close association by using either a dash or a colon.

The plumber located the problem—every one of the pipes leaked.

We can fully agree: All people are created equal.

## Complex Declarative Sentences

A complex sentence has one independent clause and one or more dependent clauses.

When the war is over, we will all enlist again.

When the war is over, if it ever is, we will all enlist again.

Although the plumber located the problem, she found it difficult to repair.

Complex sentences are simple sentences to which a qualifier (in the form of a dependent clause) has been added. Qualifications are often essential in scientific writing. The clear meaning of the simple sentence is qualified by an "if," a "when," a "whereas," an "although," or some such.

When the temperature decreases, less activity is observed.

If glucose is added, the medium will support growth.

Often, the dependent clause adds useless words, and the sentence would be stronger if the dependent clause were deleted.

Although I am not absolutely sure of what I am saying, I think that A equals B.

## Compound-Complex Declarative Sentences

A compound-complex sentence has at least two independent clauses and at least one dependent clause. Although, such sentences tend to be longer than the other three types of declarative sentences, they need not be difficult. The independent clauses are joined together exactly as they would be in a compound sentence. A dependent clause is added as it would be in a complex sentence; for example, a dependent clause at the start of a compound-complex sentence should be followed by a comma.

Although the plumber located the problem, she found it difficult to repair; in fact, she determined that new parts would be needed.

When I sell liquor, it's called bootlegging; when my patrons serve it on silver trays on Lake Shore Drive, it's called hospitality.

—Al Capone

A man who has never gone to school may steal from a freight car; if he has a university education, he may steal the whole railroad.

—Theodore Roosevelt

You can assemble and punctuate accurately all four types of declarative sentences if you learn to recognize the differences between the two types of clauses. In short, an independent clause is a complete statement; a dependent clause is incomplete, and it requires an independent clause to complete its meaning. A good test is to look at the clause and imagine that the first word starts with a capital and the last word is followed by a period. If it seems complete, it is an independent clause. If it somehow leaves you dangling, waiting for additional information, it is a dependent clause. In time, your eye and ear will recognize the distinctions, and the commas and semicolons can be inserted appropriately.

The four types of declarative sentences can be summarized as follows. And you *must* be able to construct and properly punctuate these four types of sentences because 99.44% of all the sentences you will ever write will be of these types.

A *simple* sentence is used to make an unqualified observation.

A *compound* sentence is used to make two (or more) unqualified observations, often in comparison or contrast.

A *complex* sentence adds a qualification or subordinate idea to an observation.

A *compound-complex* sentence presents two primary observations, one or both of which are qualified.

## LENGTH OF SENTENCES

Short sentences are better than long sentences, as a rule. Our eyes and our minds can grasp the meaning quickly if the words are presented in short groups.

Having said that, I now ask you to look at sentences from a somewhat different perspective. For example, does a "sentence" really have to be a full statement (with subject and verb)? No. The preceding "sentence," the word "No," would be called a "fragment," and many grammarians believe that a "frag" (which they love to mark on students' papers) is always wrong. I disagree, and I have used a number of "frags" in this book. Many writers use them to good effect.

> There's nothing to winning, really. That is, if you happen to be blessed with
> a keen eye, an agile mind, and no scruples whatsoever.
> —Alfred Hitchcock

The first of these two sentences is a full sentence. The second, although much longer, is a fragment. Together, the two "sentences" are very effective.

Short, simple declarative sentences can be logical, perhaps even pretty. But, *are they complete?* A short sentence that says it all is the ideal; however, a short sentence that omits something important is not a good sentence. Scientists, especially, must always remember the necessity of qualifications. Rarely can you write a simple sentence such as this:

> Compound A increased the survival rate by 40%.

Perhaps the survival rate was indeed increased by 40%, but over what period of time, under what conditions? So, you may need to add one or more dependent clauses and perhaps another independent clause. True, some additional material is best added in the form of additional sentences. But two sentences suggest to our mind's eye that there is a strong degree of independence between the two thoughts. When we need to show *close* relationships, it is often better to add clauses to a sentence.

> When the infected animals were treated for 3 days, compound A (100 ng per
> mg of body weight) increased the survival rate by 40%.

A short sentence is not necessarily a good sentence, and a long sentence is not necessarily bad. Short sentences *can* be confusing. Very long sentences *can* be very easy to read if their phrases and clauses are presented in logical order with proper punctuation.

Various experts have proposed readability formulas and indexes designed to measure the difficulty of written materials. I agree with their precepts that short words are better than long words and short sentences are better than long ones. However, I have never tried to reduce the number of syllables or words in my own sentences—nor in anyone else's. To me, writing by the numbers does not make good sentences any more than painting by the numbers makes good art.

The effective writer should seek to excise unneeded words from all sentences, short or long. The good writer should use the most meaningful, exact words, short or long.

---

Clearness is the most important matter in the use of words.

—Quintilian

# Chapter 14
# Paragraphs

*Paragraphs are not just chunks of text; at their best, they are logically constructed passages organized around a central idea often expressed in a topic sentence. A writer constructs, orders, and connects paragraphs as a means of guiding the reader from one topic to the next, along a logical train of thought.*

—Victoria E. McMillan

## ORGANIZING THOUGHTS

A sentence is the basic unit of communication. A paragraph consists of one or more sentences on essentially the same subject. If each set of closely related sentences is neatly packaged in its own paragraph, the reader has no difficulty in following related items; then, with the start of a new paragraph, the reader is automatically prepared for a new set of thoughts, perhaps additional information related to the preceding paragraph, or perhaps material contrasting with the preceding paragraph. If paragraphs are created with care, they are of great value to the reader in following the writer's logical development of thoughts and arguments.

## PARAGRAPH STRUCTURE

An effective paragraph should have an obvious beginning, middle, and end. The beginning (often just the first sentence) should state clearly the subject of the paragraph. The next group of sentences should marshall the evidence in support of or describing the subject. The final part of the paragraph (often the final sentence) usually provides a conclusion or a summary of what has been said in the paragraph, or it provides a transition to the next paragraph.

The first sentence (called the topic sentence) is the key. If the general subject of the paragraph is not clear, the details that follow in the body of the paragraph will hardly interest the reader.

The last sentence is usually important also, underlining the conclusion or at least the gist of the paragraph. Some paragraphs, however, must end without a definite concluding statement. When this happens, the first sentence of the *next* paragraph should make it clear that a transition in subject or development has occurred. Fortunately, English has many words and phrases that can be used as transitional signposts to aid the reader. If the new paragraph will present different but supportive information, the first sentence might start with such expressions as

> In addition,
>
> Moreover,
>
> We also noted that
>
> Smith et al. (1990) obtained somewhat similar results, showing

If the new paragraph will present contrasting information, suitable transitions might be

> On the other hand,
>
> However,
>
> Contrary to this suggestion,

If the new paragraph is to serve as a conclusion to the previous paragraph, group of paragraphs, or perhaps the entire paper, transitional openings might be

> Thus,
>
> Consequently,
>
> In conclusion,

Below is a paragraph from an article about head movement in the barn owl (T. Massino and E. I. Knudsen, *Nature* 345:434-437, 1990). Note how the topic sentence introduces "saccade generators," the subject of the paragraph. The second and third sentences go on to describe different aspects of the subject under discussion. The final sentence then neatly summarizes what has been said about the saccade generators.

> These results suggest that movement direction is encoded by the coordinated action of a small number of distinct neural circuits (referred to henceforward as saccade generators), each controlling movement in a particular direction. We determined the number of saccade generators and the directions they encode by examining the effect of one movement on another for many combinations of movement directions. Whenever the two movements were

into adjacent quadrants of space, the direction of the second movement at short intervals was altered predictably to one of four directions: upward, downward, leftward, or rightward (Fig. 3). These results suggest that the tecta command for movement accesses four independent saccade generators, each responsible for the component of a movement in one of these four orthogonal directions.

Should paragraphs be short or long? A too-long paragraph may leave the reader gasping for air. Too many short paragraphs make jerky reading. The point to remember is that each paragraph should tell a story. Some of these stories will be short, and some will be long.

---

Good writing does not come from fancy word processors or expensive typewriters or special pencils or hand-crafted quill pens. Good writing comes from good thinking.

—Ann Loring

# Chapter 15
# Voice, Person, and Tense

*For years, teachers and editors together have labeled the passive voice the bete noire of governmental and industrial writing and have conducted an intensive campaign against it. The complaints are familiar: The passive is weak, evasive, convoluting, confusing, tentative, timid, sluggish, amateurish, obscene, and immoral.*

—Don Bush

## VOICE

English has two "voices": active and passive. If the subject of the sentence is the "doer," the sentence will be in the active voice; if the subject of the sentence is the recipient of the action, the sentence will be in the passive voice.

> (active) John hit Jim.

> (passive) Jim was hit by John.

These two sentences say the same thing, but note that the active sentence has three words and the passive has five words. Passive-voice sentences are always wordier than active-voice sentences. Therefore, the guideline should be: The passive voice should be avoided. Uh-oh. That's not right. Let's try again. Avoid the passive voice.

Note that I did not say the passive voice should never be used. Use the active voice most of the time, because it is usually more direct and always less wordy. However, some sentences are *better* in the passive voice because the writer wants to emphasize the action rather than the agent of the action.

Streptomycin, the first effective cure for tuberculosis, was discovered by Selman Waksman.

Selman Waksman discovered streptomycin, the first effective cure for tuberculosis.

The first example (in the passive voice) emphasizes the antibiotic and the disease, rather than the discoverer of the antibiotic. The active voice version brings Waksman to the fore but has the effect of deemphasizing the antibiotic and the disease. If the writer is mainly concerned with tuberculosis or streptomycin, the passive version would be the better choice; if the main concern related to Waksman, the active version would be better.

Furthermore, many passive sentences really can't be converted to active.

Petri dishes are made of plastic.

A sentence such as this is passive of necessity. One does not have the choice of saying "Make petri dishes out of plastic."

## PERSON

The main reason for avoiding the passive voice is not its wordiness but its confusion as to who is doing what to whom. More often than not, the confusion results from avoidance of first-person pronouns. Much violence has been done and is being done to the English language by people whose foolish false modesty prevents them from using the personal pronouns "I" and "we." Instead of saying "I showed that substance A affected agent B," such writers hide the agent of the action and convert the sentence into the passive voice:

It was shown that substance A affected agent B.

Now we have confusion (in addition to the added wordiness). *Who* showed? *When* was it shown? (The "I" or "we" would at least limit the time to the life of the author; the "it was shown" version could refer to prehistoric times.)

Because of my years of preaching on this subject, I find it satisfying that in recent years most scientists have adopted the "first-person" style. When writing this chapter, I checked the two issues of *Science* then on my desk. Happily, the authors of 12 of the 14 Reports in the 1 June 1990 issue used the first person, and the authors of 9 of 11 Reports in the 8 June 1990 issue used the first person.

Friends, it is not egotistical to say "I" or "we." It is simply stupid not to.

## TENSE

In scientific writing, only two tenses are normally used: present and past. Occasionally, the future tense might be used, in pointing to the need for further experimentation, for example. And only rarely should one use the so-called "perfect" tenses.

Researchers *have* used this common procedure for a variety of tests.

Researchers *had* used this common procedure for a variety of tests.

When perfect tenses are used, be careful to distinguish between the past perfect and present perfect. The use of *have* in the first example indicates that the procedure is still in use. The *had* in the second example indicates that the procedure is no longer in use.

In citing references, many authors habitually write "Jones (1989) has found that . . . ." The use of simple past tense saves a word in each such instance, and the meaning is the same: "Jones (1989) found that . . . ." "It was found" is shorter and clearer than "It has been found." (But "I found" wins the prize.)

The present tense is used for established knowledge and for previously published work (including your own).

The Earth *is* round.

The first quinone to be reduced *is* tightly bound (Jones, 1989).

Knowledge has not been established, however, until *after* publication in a primary scientific journal. Thus, give your present experimental results in the past tense.

The subclones containing the D-helix substitutions *were* reassembled into plasmid P112924.

Table 4 *shows* that the L- and M-subunit genes *were* separately subcloned into M13 derivatives.

In the example immediately above, the past-tense "were" was properly used to describe present results. However, note that the present-tense "shows" was also used. This follows the rule of *presentation*. The results in a table or figure are available in the present for the reader to examine, and thus the present tense is used.

Figure 1 *indicates* that . . . .

This formula *suggests* that . . . .

Another rule relates to attribution. To show respect for established knowledge (previously published work), use the present tense; however, *attribution* is given in the past tense.

Jones (1989) *showed* that the first quinone to be reduced *is* tightly bound.

Thus, we have four "tense" rules that should normally be followed in scientific writing:

1. Established knowledge (previous results) should be given in the present tense.
2. Description of methods and results in the current paper should be in the past tense.
3. Presentation (Table 1 shows that . . . ) is given in the present tense.
4. Attribution (Jones reported that . . . ) is given in the past tense.

In the typical scientific paper, these four rules will result in frequent tense changes, often in the same sentence. But these rules should be followed carefully, particularly the rule that specifies that your present results should be in the past tense. Otherwise, the reader will have difficulty in distinguishing your new results from previously published work.

---

For one word a man is often deemed to be wise, and for one word he is often deemed to be foolish. We should be careful indeed what we say.
—Confucius

---

# Chapter 16
# Punctuation

*For most of us, punctuation is not an aesthetic challenge but a practical housekeeping problem: We engage it only long enough to keep things straight. And yet, deployed carefully and sensitively, commas, colons, and semicolons can make our sentences not only clear but even a bit stylish. Good punctuation won't turn a monotone into the Hallelujah Chorus, but a bit of care can produce gratifying results.*

—Joseph M. Williams

## THE MARKS AND THEIR MEANING

Punctuation is easy. Compared with grammar and its profusion of rules, most of them outmoded if not outright wrong, punctuation has a number of clear, simple rules. And they work. And they are easy to learn if you will but try. (These suggested rules are briefly listed in Appendix 1.)

Professor Williams (see above) says that punctuation is "housekeeping." I prefer a different metaphor, although mine doesn't have a Hallelujah Chorus. Think of words as an almost endless movement of automobiles along roads, through intersections, entering into hospital and school zones, and needing a great many signals to avoid collisions, traffic jams, and potential mishaps of many kinds. Now think of punctuation marks as a set of traffic lights and road signs, which, if well designed and well placed, will keep traffic moving smoothly. Some signs will say stop. Others will say slow down. Still others will indicate, in a variety of ways, how to drive safely through the sentences of our writing.

English has only 14 punctuation marks. If you try, you can learn the rules regarding the use of each. If you then apply these rules, your words and sentences will flow smoothly and deliver your passengers (readers) safely to their destination, the meaning of what you are writing.

66

# PERIODS

A period is used to indicate the end of every sentence that is not a question or an exclamation. Thus, all four types of declarative sentences (*see* Chapter 13) end with periods. Think of a period as a *full stop* (which is what the British wisely call it). That is really all you need to know about periods, except for a few special uses as follows.

Periods are also used after some abbreviations. (Style regarding the use of periods after abbreviations varies among journals.)

> R.A. Day
>
> Feb.

Periods are used in decimals.

> 0.4
>
> 3.1416
>
> $4.98

Periods have two additional purposes. In quoted material, three periods (ellipsis marks) are used to indicate that material has been omitted. If the omission occurs at the end of a sentence, four periods are used (three ellipsis marks and one period ending the sentence).

> He said "Please pass the . . . potatoes."
>
> She answered "Pass your own. . . ."

Another use of periods is as "leaders," especially in lists and in tables, to help the reader.

> Penicillin ............................... 5 g
>
> Dihydrostreptomycin. ............ 2 g

# QUESTION MARKS

Question marks are easy. Place one at the end of any sentence that asks a question.

> What time is it?

Some sentences look like questions, but in fact are not questions.

> Would you please turn in the assignment that is due today.

This sentence is in the form of a question, but it is really an order; a period is therefore the appropriate punctuation mark.

He asked me whether I was going.

This sentence is an indirect question. Such "questions" are followed by periods rather than question marks.

Unlike periods, a question mark can be used at the end of a parenthetical clause within a declarative sentence.

On Monday (or was it Tuesday?), I completed the experiment.

## EXCLAMATION POINTS

An exclamation point follows any word, group of words, or sentence that expresses excitement, fear, surprise, or alarm.

Ouch!

Hot stuff!

Get out of here!

Exclamation points are rarely used in scientific and professional writing. In fact, they should be used with relative rarity in any kind of writing, or they lose their force.

What a day!

What a day.

Which example is correct? It depends on the "force" the writer intends. Exclamation points, like the other punctuational road signs, communicate the writer's intent. In the first example, if the "day" referred to had brought a tornado or an earthquake, the sentence certainly should end with an exclamation point. If the day were simply a busy day, however, the mild exclamation (ending with a period) would be appropriate.

Damn!

Jeepers.

The same basic principle operates here, the first example showing a real expletive, the second showing a mild oath.

## COMMAS

Periods, question marks, and exclamation points are used to signal the ends of sentences. All of these marks clearly tell the reader: "Stop. This sentence is over."

For those of you who don't know a comma from a long sleep, I will define the comma as a road sign that says "Slow down." Commas are used within sentences, never at the end. Commas have many uses within sentences, but the signal always has the same intent: slow down, pause, take a breath. (How is a cat different from a comma? A cat has claws at the end of its paws; a comma is the pause at the end of a clause.) People who don't know how to use commas tend to haphazardly sprinkle them over the page (as in the *Peanuts* cartoon).

Reprinted by permission of UFS, Inc.

A common use of the comma is to separate the clauses in a compound sentence. Such usage (see Chapter 13) requires a coordinating conjunction in addition to the comma.

A woman drove me to drink, and I never even had the courtesy to thank her.
—W. C. Fields

Many authorities say that a comma is not necessary between independent clauses if both are short.

Who is up and who is down?

I would say, however, that you should get into the habit of using the comma for this purpose. *It is never wrong.* And then you won't have to keep asking yourself whether the clauses are short enough to be attached without a comma.

Another common use of commas is to signal the end of an introductory word, phrase, or clause and the start of the main clause.

Hey, cut it out!

In particular, commas are important.

On the whole, holistic medicine is a wholly new subject.

When Irish eyes are smiling, sure they'll steal your heart away.

Again, many authorities say that a *short* introductory phrase or clause need not be followed by a comma. However, look at the second example above. "In particular" is certainly a short phrase, but, without a following comma, the reader cannot quickly make sense out of "In particular commas." Again, I say use a comma after every introductory word, phrase, or clause. The road sign will usually be helpful to the reader.

Unfortunately, this rule does not hold for phrases and dependent clauses that *follow* the main clause. A comma is sometimes used, sometimes not, depending on whether the writer wants the reader to slow down.

> You will be visited by the first of these spirits when the bell tolls one.

> You will hear the bell, if it rings.

Commas are also used to set off interrupting elements in the middle of a sentence.

> This book, I think, may be the dullest ever written.

Note that the interrupting element is preceded by a comma and followed by a comma. Both commas (called balancing commas) are necessary, and they have the effect of taking the enclosed material out of the main sentence. Usually, parentheses could be used to replace the commas.

Balancing commas are also used around appositives (as in the first example below). Incidental (parenthetical) information within an independent clause can also be set off with commas.

> His name, Cholmondelay, is very British.

> Day, one of the great minds of the preceding century, can't write worth a nickel.

We could almost say that any phrase or dependent clause placed in the middle of an independent clause should be set off with balancing commas. Rarely, however, an intervening phrase or clause is essential to the meaning of the independent clause; such elements are not "parenthetical" and commas are not used.

> Calculus, like many subjects, requires much study.

> People like Jimmy Bakker make me ill.

These examples are identical in construction, but the "like" prepositional phrase in the first is parenthetical (not essential to the meaning of the statement), whereas the "like" phrase in the second is essential. (Without it, the sentence would be "People make me ill," which would be a serious malady indeed.)

Three or more words or groups of words in a series are separated by commas.

> She published novels, poems, and short stories.

> His books were admired by every Tom, Dick, and Mary.

The big argument is whether a comma should be used before the "and" (or "or") in the series. Here is another rule. *Always* use the comma before the "and." *It is never wrong.* People who never use the series comma, or who use it only selectively, will write many sentences that are hard to read and a few that are incomprehensible. If you use these commas to separate the three parts of a series, you will never write a sentence like this:

> He had a large head, a thick chest holding a strong heart and big feet.

The only way to get the feet out of that poor guy's chest is to put a comma before the "and."

> The system consists of an engine, tubing to bring fuel to the cylinders and associated mounting bolts.

This sentence is incomprehensible, like many constructed by people who fail to use serial commas. Possibly, the tubing brings fuel to the mounting bolts. (If so, the sentence would be improved by putting an "and" before "tubing.") More likely, the tubing takes fuel to the cylinders and *not* to the bolts. If so, a comma intelligently placed after "cylinders" clears up all confusion.

Commas are sometimes used to separate adjectives that modify the same noun, especially, when three or more adjectives are involved. With two adjectives, the comma is used if the adjectives are totally unrelated to each other.

> It was a dark, cold, stormy night.

> It was a dark, stormy night.

> It was a dark green dress.

In the third example, the dress is "dark green"; therefore, "dark" is not separated from "green." (Because "dark" modifies "green" rather than "dress," many writers would use a hyphenated expression: "dark-green dress.")

Whenever a writer wants the reader to pause for a moment, perhaps for emphasis or contrast, a comma is the appropriate road sign.

> He went into the saloon, again.

Commas are used to set off small units from larger ones.

> Newark, DE, U.S.A.

Commas are used to set off names in direct address.

Go to it, John.

Commas are used to separate numbers (but not numbers within a year).

1,000

10,000,000

1990

Use commas before degrees and titles.

Mary Jones, Ph.D.

Nancy Kassebaum, U.S. Senator

In a sentence, such degrees and titles give incidental information, so balancing commas are used.

Mary Jones, Ph.D., is the project leader.

A comma, rather than an exclamation point, is used after a weak exclamation.

Well, so it goes.

A comma can be used to indicate the omission of words in a sentence.

She went to the right; I, to the left.

Finally, commas may *not* be used to separate a subject from its verb or a verb from its object, unless a balancing comma is used.

I an immunologist, tested the serum samples.

I, an immunologist, tested the serum samples.

The first example is wrong because the subject ("I") is separated from the verb ("tested"). The second example is O.K., because the balancing commas effectively take the appositive ("an immunologist") out of the sentence.

## SEMICOLONS

Once you have learned that a semicolon is something other than half a colon, you will find that semicolons are easy to use. Their main purpose is to separate the independent clauses in a compound sentence. A comma will suffice if a coordinating conjunction accompanies it. Often, however, the writer may want to indicate the equivalence of the clauses by directly welding them together with a semicolon.

> Lead me not into temptation; I can find the way myself.
> —Rita Mae Brown

> Hanging is too good for a man who makes puns; he should be drawn and quoted.                              —Fred Allen

> In this world of sin and sorrow there is always something to be thankful for; as for me, I rejoice that I am not a Republican.
> —H. L. Mencken

A conjunctive adverb may be placed after the semicolon (and a comma must follow the adverb).

> Television is a popular medium; however, I would define it as the bland leading the bland.

The only other use for the semicolon is to separate the parts of a series that has one or more internal commas. A semicolon should precede the "and" in the series, just as the comma precedes the "and" in a simple series.

> On the table were oranges from Florida; pears from Washington, Oregon, and California; and apples from Oregon.

## COLONS

A colon is used to introduce a word, a phrase, a clause, or a sentence.

> There was only one appropriate color: pink.

> Chastity: the most unnatural of the sexual perversions.
> —Aldous Huxley

> Only one thing counts: Is it true?

The colon often introduces a list. However, words like "the following" should be used to signal to the reader that a list is coming.

> My shopping list included: apples and oranges.

> My shopping list included the following: apples and oranges.

The colon is used erroneously in the first example, but it is used properly in the second. Grammatically, the first sentence is incorrect because the colon separates a verb from its object. In the second example, "the following" becomes the object, and "apples and oranges" becomes an appositive.

Rarely, a colon is used in place of a semicolon to separate two independent clauses. This is done when the second clause has a strong relationship to the first.

> Publicity is like poison: it doesn't hurt unless you swallow it.
>
> —Joe Paterno

A semicolon would not be incorrect in this example, but the colon does a better job of unifying the two clauses.

Colons are often used to separate a book, chapter, or article title from a subtitle; a semicolon, though often used in these instances, is incorrect.

> Advertising: The World's Oldest Profession

> Scientific English: A Guide for Scientists and Other Professionals

The only other uses of the colon are certain conventional uses, such as at the end of salutations in letters, in between hours and minutes, and in ratios.

> Dear Sir or Madam:

> 2:30 A.M.

> 1:3

> 1:3::2:6

# DASHES

Dashes are overused by many writers, especially inexperienced writers. Writers who do not understand the niceties of balancing commas and parentheses are likely to dash their sentences to death. Nonetheless, there are four uses of the grammatical dash. Dashes can be used to set off appositive or contrasting information.

> There was one important man in her life—her father.

> In the game of life, there are no winners—only losers.

Singly or in pairs, dashes can be used to set off words that summarize or provide examples.

> From reading this book, you are learning material—grammar, spelling, punctuation—that will always be useful to you.

> He's the kind of man who picks his friends—to pieces.
>
> —Mae West

> The sound of a harpsichord—two skeletons copulating on a tin roof in a thunderstorm.
>
> —Sir Thomas Beecham

A dash, like a colon or semicolon, can be used to link two independent clauses. A dash would be a good choice if the second clause provides a surprise or contrast.

She smiled when the thief ran off with her handbag—it was empty!

I don't know anything about music—in my line you don't have to.
—Elvis Presley

Finally, a dash can be used to set off interrupting words.

He is a smart fellow—or so I am told—but nobody believes what he says.

A dash is, of course, different from a hyphen. Some typewriters and word-processing keyboards have a hyphen but not a dash. In this situation, a single hyphen means a hyphen, and a dash is indicated by two hyphens. There should be no space before or after either hyphens or dashes.

## QUOTATION MARKS

Whenever you are directly quoting someone else, use quotation marks. In the U.S., double quotation marks are used. Single quotation marks are used to indicate quotations within quotations. In the United Kingdom, usage is considerably different, both as to the type of quotation marks used and as to placement of other punctuation marks within or outside closing quotation marks. With apologies, I will present American usage only. Because quote marks are the most difficult of all punctuation marks, the less complexity the better.

He said "I love you."

"I love you," he said.

"I said to my boyfriend Ernie 'Ya gotta kiss me where it smells.' So he drove me to Wapping."

—Bette Midler

Note the placement of the period in the first example and the comma in the second. You may now memorize Quotation Rule 1: Commas and periods *always* go inside closing quote marks.

He said "I love you"; she said "Get lost."

"I love you": those were his words.

These examples illustrate Quotation Rule 2: Colons and semicolons *always* go outside closing quote marks.

Question marks and exclamation points are placed inside the quote marks if the person quoted is asking the question or exclaiming.

He asked, "When are we going?"

She shouted "Fire!"

Question marks and exclamation points are placed outside the quote marks if the writer is asking or exclaiming.

> Did he say, "The rent is due"?
>
> When offered the money, she said "No, sir"!

Because it would be surprising for anyone to say no to the offer of money, the exclamation point in the second example is appropriate, and it is placed correctly.

If both the writer and the person being quoted are asking questions or exclaiming, the marks go inside the closing quote marks.

> Did he ask, "When are we going?"
>
> In terror, I yelled "Get out of here!"

Quotation Rule 3 thus states that question marks and exclamation points go either inside or outside closing quote marks. If the quoted person is asking or exclaiming, the marks go inside; if the author is asking the question or exclaiming, the marks go outside; if both the writer and the quoted person are asking or exclaiming, the marks go inside the closing quotes.

> The lion roared: "Who's King of the Jungle?" The gazelle answered "You, oh King!" The lion roared again: "Who's King?" The monkey answered "You, of course!" Once more the lion roared, this time right into the face of the elephant. Whereupon the elephant picked up the lion in his trunk, slammed him on the ground several times, and threw him in a heap under a tree. The lion said "Hey, you don't have to get mad just because you didn't understand the question!"

If you quote a long passage, you should begin each new paragraph with quote marks. Closing quote marks should be placed *only* at the end of the final paragraph. But: if you indent quoted material, or have it set in small type, you should not use quotation marks at all.

*Never alter a quotation.* If you omit material from a quotation, indicate this with ellipsis points (three periods). If you add anything to a quotation, the added material should appear within brackets. You should not even correct a misspelled word if one occurs in a quotation. (It is often wise to put *sic* within brackets after the misspelling to indicate that the misspelling appears in the original.) Use brackets to change pronouns or verb tenses if necessary. Examine the following sentence:

> He boasted that he "can do twice as much work as I can."

Almost certainly, the person quoted must have said "I can do twice as much work as you can." The example above should either be paraphrased (by eliminating the quote marks) or rewritten with bracketed insertions.

He boasted that he can do twice as much work as I can.

He boasted that "[he] can do twice as much work as [I] can."

Quotation marks are conventionally used to enclose titles of poems, articles, stories, and chapters of books. (Quote marks should not be used for titles of books or journals; these should be italicized.)

Words that are used as words can be in italics or within quote marks.

The word *computer* means many things.

The word "computer" means many things.

## APOSTROPHES

Apostrophes are used to show possession. They must be placed carefully.

The boy's dog

The boys' dog

Note the different meanings in these examples. In the first example, one boy has a dog; in the second example, two or more boys have a dog.

People worry about when to add an "s" to the apostrophe when forming possessives. In the singular, add the apostrophe *and* the "s" in almost all instances.

Day's Rule

Jones's Rule

Weiss's Rule

Even with the tripling of the "s" in the third example, this guideline can be defended. In the plural, add an apostrophe and "s" to a word that does not end in "s."

We sell men's clothing.

If a word ends in "s," add only an apostrophe.

Professors' salaries are low.

Apostrophes are also used in contractions; however, contractions should seldom be used in scientific writing.

I can't do it.

It's in the bag.

If my aunt had wheels, she'd be a trolley car.

In the first example, "can't" is a contraction of "can not"; in the second example, "It's" is a contraction of "It is." Unfortunately, "it's" is often confused with the possessive pronoun "its," which does not have an apostrophe. Get into the habit of looking at your "its." When the "its" means "it is," put the apostrophe in its (not it's) place.

> Its important to use apostrophe's correctly.

In the above example, *Its* should be *It's* (contraction of *It is*) and *apostrophe's* should not have an apostrophe (because *apostrophes* is a simple plural).

© 1991 by Daniel Hudgins.

Apostrophes can sometimes be used to form plurals. (Usage varies from journal to journal and publisher to publisher.)

> three 4's
>
> two a's
>
> several DNA's
>
> the 1980's

Many publishers insist, however, that scientific abbreviations are singulars, not plurals. Thus, you should not use "several DNA's" or "several pH's" (with or without the apostrophes). Instead, you should refer to "several DNA preparations" and "several pH values."

## PARENTHESES

Parentheses can be used around words, groups of words, or whole sentences that add incidental information.

Some people (Einstein, for example) are smarter than others.

If a parenthetical element appears at the end of a sentence, the period goes outside the closing parenthesis.

He went into the bar for a quick one (which became a quick three).

A complete sentence within parentheses starts with a capital and ends with a period; the period should be inside the closing parenthesis mark.

He went into the bar for a quick one. (She did a slow burn.)

However, if the parenthetical sentence is within another sentence, a different rule applies.

Some parenthetical sentences (note the punctuation of this one) need neither a capital nor a period.

## BRACKETS

Rarely, you may need to give additional information within material that is already in parentheses. Brackets can be used for this purpose.

(The cost was 3 pounds [about 6 dollars].)

The main use for brackets is to enclose alterations or additions to quoted material.

"But when he [George Washington] was elected President, we forgot the cherry tree."

"He was verry [sic] sad."

## SLASHES

The slash (solidus, virgule, shilling) is considered to be a mark of punctuation. In my opinion, it should not be so used. It can be used as a mark of division.

$4/2 = 2$

But, precisely because it means "four divided by two," it should not be used grammatically. People who use "he/she" certainly do not mean he divided by she. People who use "noise/signal ratio" usually mean "signal to noise ratio." People who use "and/or" do not know what they mean.

## HYPHENS

Ah, yes, the hyphen. The hyphen has a number of uses, most of them confusing.

Hyphens are used to indicate end-of-line word division in printed materials. This should not be a problem; any good dictionary will indicate where a word may be broken (between the syllables, of course). However, I should warn you that you should proofread typeset materials carefully. Today, we rely on computers to break the words. In the old days, we trusted human beings, the Linotype and Monotype operators, to perform this function. The computer program has not yet been written that can perform this function with the accuracy of the old human typesetters. Look at the line breaks in almost any newspaper or other publication today, and you will see what I mean. Every bad break will slow down the reader, and some bad breaks lead to total confusion.

So, check a dictionary for proper placement of end-of-line hyphens. But also check your head. The dictionary may say that *analogy* is divided *anal-o-gy;* however, if it comes out *anal-ogy* in your paper, your readers could be confused. I once saw a line break that yielded *fig-urine* (which is right according to dictionaries); my reading concentration was totally broken. Because many words in English can be confusingly divided, proofread carefully and make sure that words are divided sensibly.

A second use of hyphens is to link the parts of certain compound words. When the plural is formed, only the significant part of the term takes an "s."

| | |
|---|---|
| mother-in-law | mothers-in-law |
| attorney-general | attorneys-general |
| brother-in-law | brothers-in-law |

I want to be the white man's brother, not his brother-in-law.

—Martin Luther King, Jr.

And now we come to what is sometimes called the unit hyphen. When two words modify a third, it is sometimes necessary to link them with hyphens to indicate that they are acting together.

I don't like Smith's third-rate book.

In this example, the unit hyphen must be used to link "third" and "rate." It is not Smith's "third book." (It may be her first.) It is not a "rate book." Thus, we must have the hyphen to show that it is a "third-rate book."

> a light red tie
>
> a light, red tie
>
> a light-red tie

Of these three examples, the first is the least satisfactory. Is the tie a "light" tie (perhaps a lightweight fabric) or is it "light red"? The second example makes it clear that the tie is both "light" and "red." The unit hyphen in the third example makes it clear that the tie is "light red."

> a new car owner
>
> a new-car owner
>
> a new car-owner
>
> imported-car dealer

Again, the first of these three examples is imprecise because we don't know what is "new." The second example is clear: the owner has a new car. The third example is also clear: we have a new "car-owner." (Presumably, this person has not previously owned a car.) The fourth example is correct with the hyphen; nobody would import car dealers.

The unit hyphen is sometimes clearly needed.

> I am easily satisfied: an easy chair, a good book in one hand, and a single-malt whisky in the other.

The hyphen is mandatory. I certainly don't mean "a single whisky" (because I usually have a double); I certainly do mean "single-malt" (because they are so much better than the cheap blends).

In scientific writing, the unit hyphen is used many, many times. It should be used whenever two words are linked together to modify a third word. Some common examples: riboflavin-binding protein, penicillin-resistant streptococcus, gram-negative bacteria.

Finally, hyphens are sometimes used to join prefixes and suffixes to words. Although this has been a troublesome area in the past, modern usage is tending strongly toward eliminating such hyphens. You should now get into the habit of linking all standard prefixes (e.g., *re, pre, non*) to their following words, without a hyphen. Never again use "non-returnable"; use "nonreturnable."

Some authorities have said O.K., let's eliminate the hyphen, but only if it does not involve doubling of the vowels. For example, they would accept "retest" and "predetermine," but they would insist on "re-elect" and "pre-eminent." Most modern writers now reject such nitpicking, and they "reelect" "preeminent" candidates to their heart's content.

So join the joiners. About the only exceptions you need to watch for are prefixes followed by proper nouns or numbers. These still take hyphens.

pre-Sputnik

pro-Democratic

pre-1990

Suffixes are also joined to the root word as a rule. However, if this brings about doubling or tripling of a letter, or if the word is very long, the hyphen should be used.

lifelike

tinsel-like

ball-like

pleuropneumonia-like organism

For those of you who have found this chapter to be heavy going, there is still hope. Perhaps everything will become clear when you read Appendix 1, "Principles of Punctuation Presented Plainly."

---

If you take the hyphen seriously, you will surely go mad.

—John Benbow

# Chapter 17
# Redundancies and
# Jargon

*The air is full of redundancies: My life is clogged with them. I see them in print and hear them in speech and have them spoken to me in airliners, where I am told that I now can take with me a "free complimentary" copy of their in-flight magazine. I am so grateful.*

—Richard Cohen

## DOUBLESPEAK

Redundancies come in different varieties. One common redundancy is the simple doubling of words that have the same meaning. For example, in the above quotation, either "free" or "complimentary" should have been deleted, because they both mean the same. A close relative is "free gift." Even worse is a "free, complimentary gift," a double redundancy. Another kind of doubling would make a fine bumper strip:

Help stamp out and abolish redundancy.

## USELESS WORDS

Many of the words we use are useless, but, by habit, we go on using them. We think we are impressing people. Here are some examples, the italicized words being useless or of questionable value.

| | |
|---|---|
| *active* consideration | *light* snack |
| *alternative* choice | mix *together* |
| *closely* scrutinize | *past* experience |

83

| | |
|---|---|
| *complete* stop | *past* history |
| *completely* accurate | *perfectly* clear |
| *component* part | *personal* belongings |
| consensus *of opinion* | *present* incumbent |
| *definite* decision | *qualified* expert |
| *exact* same | *securely* fastened |
| *firmly* commit | *terrible* disaster |
| *fully* recognize | *totally* useless |
| *grave* emergency | *truly* significant |
| *immediate* vicinity | *unfilled* vacancy |
| join *together* | *utterly* unique |

The above list contains two-word combinations, with one word in each pair being redundant. Even worse are whole constellations of words that roll off the tongue or the pen, most or all of the words adding nothing of substance.

> The trend seems to suggest that perhaps A affects B.
>
> In the present paper, the authors show that A affected B at 37 C.
>
> In the second part of the paper, it is shown that A affected B at 37 C only in the presence of exceedingly small concentrations of protease.
>
> The paper concludes with a summary of the evidence indicating that A may be, under carefully limited circumstances, an effective agent against infections caused by gram-negative organisms.

The above examples include phrases that abound in the scientific literature. What do they mean? In the first example, "the trend seems to suggest that perhaps" means nothing more than "perhaps." In the second example, common in Introductions, "In the present paper" (or "In this investigation," etc.) is usually wasted breath, and constructions like "the authors show" (or, worse, "the authors will show") add little more. In the absence of a literature citation (and a statement in the present tense), it should be *understood* that *we* (not "the authors") did it and that we are reporting it *in this paper.* The third example, a typical transitional sentence, is even worse. "In the second part of the paper, it is shown that" can be deleted. (In fact, it can be *totally* deleted.) The important thing is that "A affected B." Certainly, we don't need the "it was shown"; obviously, the authors of the paper (*we*) showed it. I believe that the passive-voice "it was shown" kind of phraseology should *never* be used. Better is "we showed that"; best is "A affected B," with it being *understood* that the authors of the present paper *showed* it. If Smith showed it, "it was shown by Smith (1980) that A affects B" is a poor substitute for "A affects B (Smith, 1980)."

The worst example of all is the final one. If any part of a scientific paper should be written with straightforward clarity, obviously it is the conclusions. To say that "The paper concludes with a summary of the evidence" is to state the obvious. Everybody knows that. And how about the rest of the sentence? Any conclusion containing the verb "may be" is not much of a conclusion, especially if it is immediately qualified by (unstated) "carefully limited circumstances."

"It's like my grand-dad always said:  'Son,' he'd say, 'if it ain't dysfunctional, don't attempt pre-emptive maintenance intervention on it.'  Or something like that."

Cartoon by Bradford Veley.

Sometimes, redundant words in a sentence are not only useless; they may add confusion. For example, I saw a sentence in an ad for a ballpoint pen that read "Did you know that 8 of 10 people use a ballpoint pen to write with?" My question: What do the other people use their ballpoint pens for?

## OXYMORONS

Sometimes two words of opposite meaning are put together. I was really taken aback when one of my students told me that he was "clearly confused" by one of my lectures.

Oxymorons are contradictory expressions. One might say that such expressions are *ambivalent,* but I am of two minds about that. Often, the joined words have opposite meanings. An example is "jumbo shrimp," "jumbo" meaning large and "shrimp" meaning small. Some of these word combinations make sense. ("Jumbo shrimp" makes sense, presumably meaning "large shrimp." On the other hand, "shrimp scampi" makes no sense; "scampi" is the Italian word for shrimp, so "shrimp scampi" can only mean "shrimp shrimp," which is moronic if not oxymoronic.) Here are a few oxymorons I like; I leave it to you as to whether they make sense:

| | |
|---|---|
| academic salary | minimal medium |
| business ethics | only choice |
| common sense | pretty ugly |
| creation science | Reagan Library |
| death benefits | rock music |
| discovered missing | slightly infected |
| dry martini | standard deviation |
| graduate student | strongly suggests |
| ill health | tax return |
| important trivia | tentative conclusion |
| Justice Rehnquist | thoroughly inadequate |
| may certainly | too few |
| military intelligence | understanding editor |

### WORDS AND EXPRESSIONS TO AVOID

Some words should be avoided because shorter, simpler words are available. Many expressions should be avoided because they are too wordy. There is nothing wrong with them, and all of us use such phrases on occasion. But habitual use of such expressions adds clutter and reduces clarity. Many of these "heavy breathing" expressions are listed in Appendix 3. Here are just a few examples; the left-hand column gives the wordy expressions and the right-hand column gives suggested replacements:

| | |
|---|---|
| a considerable amount of | much |
| a majority of | most |
| a number of | many |
| are of the same opinion | agree |
| at this point in time | now |
| based on the fact that | because |
| despite the fact that | although |
| due to the fact that | because |
| first of all | first |
| for the purpose of | for |

| | |
|---|---|
| has the capability of | can |
| in many cases | often |
| in my opinion it is not an unjustifiable assumption that | I think |
| in order to | to |
| in the event that | if |
| it is worth pointing out in this context that | note that |
| it may, however, be noted that | but |
| lacked the ability to | couldn't |
| needless to say | (leave out, and consider leaving out whatever follows it) |
| of great theoretical and practical importance | useful |
| on a daily basis | daily |
| perform | do |
| red in color | red |
| take into consideration | consider |
| the question as to whether | whether |
| through the use of | by |
| with a view to | to |

Additional redundancy (redundant redundancy?) results when "heavy breathing" occurs in a number of successive sentences.

The framistan is red in color.

The framistan is round in shape.

Also, as is well known, the framistan is an object that is made out of wood.

If we take out the heavy breathing and combine the sentences, the residue is much shorter and more digestible.

The framistan is a red, round, wooden object.

## BUZZWORDS

An especially annoying kind of "heavy breathing" is the use of fancy, fad words. Many writers mimic each other, like lemmings on the way to the sea. I will not try to list the current buzzwords because many of them are so palpably stupid that they will soon drop out of the language anyway (to be replaced by others that are equally bad). By the way, the word *buzzword* is itself a buzzword.

But I will give you a few examples. Then perhaps you can build your own list and contribute to the literature and to sanity by avoiding them.

Let me first articulate a buzzword that seems to rain down on me every day: *articulate.* Few people can state goals these days; most people *articulate* them. And have you noticed that people who *articulate goals* almost always have *overarching goals?* And they love to *address* the root causes that *impact* our society today.

Among the dumbest buzzwords these days is *jump-start.* I see this everywhere, especially in the sports pages. Sports columnists are almost certainly the most proficient buzzers, although social and behavioral scientists try hard to keep up. Sports writers buzz along like this:

> Right after the *opening gun,* Charles Barkley tried to *jump-start* the Sixers. He took a *feed* from Green, *bulled* his way *into the paint, posted up* Jordan, and *fired a shot* from *the key.* The shot *clanged off the rim,* but Charles got his own *carom* and *slam-dunked.*

In buzz language, *jump-start* is supposed to mean "quick start." But what is a *jump-start* really? On a cold day in February, your car won't start. You drag your neighbor away from his morning coffee, get him to maneuver his car next to yours, connect your battery to his with jumper cables, and then you *jump-start* your car. The whole process takes about 45 minutes and it's the slowest start you will have all year.

Scientific papers should not be written in buzzwords, slang, or any other words that are either meaningless or ephemeral. Try to write for readers in all countries in all eras.

## HACKNEYED EXPRESSIONS

Even worse than today's buzzwords and phrases are yesterday's. An expression, especially an apt metaphor, might be useful when it is new. But stale expressions are boring at best and annoying at worst. Fortunately, such expressions drop out of the language after we all get tired of saying or writing them. Nobody today is likely to refer to "the whole ball of wax," whereas almost everybody used the expression a few years ago.

Look for dated, stale expressions in your writing. When you notice them, delete them and replace them with newer or different expressions. Alternatively, give a twist to the hackneyed phrase, to add interest rather than boredom to your writing. Instead of saying "Every Tom, Dick, and Harry," try "Every Tom, Dick, and Mary." Instead of "The greatest thing since sliced bread," try "The greatest thing since buttered muffins." On the other hand, it is far better to "call a spade a spade" than to call it a "manual earth-restructuring implement." Strictly speaking, however, it is best to avoid most idiomatic expressions in scientific writing.

In order to help stamp out and abolish redundancy, needless duplication, and repetition, I ask your active consideration in determining an alternative choice to closely scrutinize your position and to arrive at a full and complete stop. To be completely accurate, we need to establish a consensus of opinion to determine the exact same conditions and firmly commit, or fully recognize, the immediate vicinity of the range of attitudes. If we join together or mix together our past experiences, past history will make it perfectly clear that light snacks are not the problem. Students should be perceived as qualified experts when it comes to their own personal belongings. It is totally useless to consider that present incumbents are securely fastened to their present attitudes. It would be truly significant, or at least utterly unique, to think otherwise.

*The Ultimate Redundant Paragraph*
Adapted from the examples listed in Chapter 17 of
Robert A. Day's *Scientific English:*
*A Guide for Scientists and Other Professionals*
**Jacques J. Pène**
School of Life and Health Sciences,
University of Delaware, Newark, DE 19716

# Chapter 18
# Abbreviations and
# Acronyms

*If there is any doubt, write the term out. Otherwise, your reader may be
in the position of the farmer who shot a crow and read the tag on his leg
that said "Wash. Biol. Surv." The farmer remarked that he washed the
crow, boiled it, and served it, but it still tasted awful. If there is any doubt,
write the term out.*

—Deborah C. Andrews and Margaret D. Blickle

## DEFINITIONS

To abbreviate means "to make briefer; *esp*: to reduce to a shorter form
intended to stand for the whole" (*Webster's Ninth New Collegiate Dictionary*). An acronym is an abbreviation of a special type that abounds in
technical and scientific writing. An acronym is "a word (as *radar* or *snafu*)
formed from the initial letter or letters of each of the successive parts of a
compound term" (*Webster's Ninth*). Thus, strictly speaking, pronounceable
"words," such as BIOSIS (BioSciences Information Service), which is
pronounced "bi-ó-sis," are acronyms, whereas abbreviations such as DNA
(deoxyribonucleic acid) are not. (You don't try to pronounce "d-na"; you
spell out "D," "N," "A.")

Acronyms and abbreviations formed by that first-letter method first
came into general use during World War II, as one might guess from the
examples used in *Webster's*. Radar (*ra*dio *d*etecting *a*nd *r*anging) came into
use around 1941, replacing a system of aircraft control known as CNS
(*c*ontrol *n*et *s*ystem; an abbreviation, not an acronym). Later in World War
II, sonar (*so*und *na*vigation *r*anging) also came into use.

Perhaps the most popular acronym that came out of World War II was *snafu,* which spawned a large progeny of related acronyms that almost always contained an "f" word. When asked by mothers, GI's would explain (at least I did) that the "f" stood for "fouled." Thus, *snafu* meant "situation normal, all fouled up." One of my favorites, *fubar,* meant "fouled up beyond all recognition."

## HOW TO ABBREVIATE

The "how" is easy. In science, clear-cut conventions have been established. First, the abbreviation is introduced at *first use* in the text. Second, the abbreviation is "introduced" within parentheses immediately after the spelled-out word or term that it abbreviates. Third, the abbreviation is *normally* given in capital letters (except units of measure), without periods and without spaces.

The reaction proceeded at standard temperature and pressure (STP).

The virus was a close relative of tobacco mosaic virus (TMV).

## WHAT TO ABBREVIATE

The *what* is not easy. In deciding whether to abbreviate a word or term, you should ask yourself three questions.

First, is there a *standard* abbreviation readily available? If so, use the abbreviation. (Some journals allow a few well-known, standard abbreviations to be used without introduction.) You may need to check the Instructions to Authors of your target journal or the principal style manual in your field.

Second, is the word or term long or short? A word like *acid* is obviously too short to merit abbreviation. (And the abbreviation A would be confusing.) A term such as *nitric acid* is still too short to merit abbreviation. (And NA would mean "not applicable.") However, when we get to a mouthful like *deoxyribonucleic acid,* we have a term that cries out for abbreviation.

Third, how often is the word or term used in the paper? If the word or term is used frequently, use the abbreviation. If it is used rarely, do *not* use an abbreviation. If an abbreviation (unless it is a standard one) is introduced in the Introduction and then not used again until the Discussion, it is likely that the reader will have forgotten what it means.

In addition, beware of introducing an abbreviation that might cause confusion. I don't think I would abbreviate *serum oxalacetic bicarbonate,* because the obvious abbreviation might be disconcerting. Some authors use

BSA for bovine serum albumin, but I then get the notion that they are writing about the Boy Scouts of America.

## WHAT ABBREVIATIONS NEED NO INTRODUCTION

Units of measure do not have to be spelled out at first use. The standard metric units are known throughout the world (among scientists at least). The abbreviations for these units are usually in lower case, without periods and without spaces.

I added 4 ml of distilled water to the reaction mixture.

Does that mean that units of measure are always abbreviated? No. They are abbreviated *only* when used with numbers. Furthermore, if they are used with a spelled-out number (at the start of a sentence), they are spelled out.

I added a few milliliters of acetone.

Four milliliters of acetone were added.

Incidentally, to avoid starting a sentence with a numeral or with a spelled-out number and unit, simply get the real subject of the sentence up front.

Acetone (4 ml) was added.

Chemical elements and compounds, like metric units, are known throughout the world, and these "abbreviations" need not be introduced. However, it is still wise to follow the "short" rule. It makes sense to use HCl in place of hydrochloric acid, but it isn't wise to use $H_2O$ in place of water (except in a formula involving waters of hydration).

---

Some authors use abbreviations freely (and coin new abbreviations) because they are too lazy to write out full terms or to pause to decide whether an abbreviation is truly needed in a particular sentence.

—Edward J. Huth

---

# Chapter 19
# Language Sensitivities

*Sticks and stones can break our bones, but names will never hurt us.*

—Old Proverb

## MINORITIES

Unfortunately, names *can* hurt us. Racial epithets, comments about our personal appearance, and insults to our religions or our beliefs can be devastating. Therefore, we must learn not only how language can be used but how it can be abused.

To coin a phrase: The majority of us are minorities. In fact, all of us are different from the majority of people in one way or another. Still, there are certain "minorities" which have been treated badly by the "majority." Some of this bad treatment was historical, but unfortunately we are still many kilometers away from a society in which all of us are "equal."

But we are improving. And language, it seems to me, leads the way. Until we can "talk a good game," resolution of problems is difficult.

In the United States, at least, substantial progress has occurred in recent years. These days, only a rube would tell an ethnic joke. And the words that formerly were used to tell such "jokes" (nigger, kike, Dago, Polack, etc.) are totally avoided by sophisticated people.

Unfortunately, some of the historical baggage we all carry is more subtle. Unconsciously, if we are not on guard, we might use a racist, sexist, or other insulting expression without realizing it.

On an exam in Technical Writing, I asked my students to correct any errors in this sentence: I gave him a box of flesh colored band aids. Some students correctly changed band aids to Band-Aids (a registered trademark).

A few students correctly placed a hyphen between "flesh" and "colored" to show that together they serve as a unit modifier. Sadly, none of the students questioned "flesh colored." What color is flesh? In most of the world, it is *not* white. For whites to go on assuming that their color is somehow the standard one is an egregious way of thinking, speaking, and writing.

## SEXIST WRITING

For centuries, in small ways and large, men have subjugated women. This fact has been strongly embedded in our language. Only in the past generation has there been a strong movement to end male chauvinism.

The most common problem in English was the use of male pronouns to mean both males and females. Because English has no singular pronoun that includes both (like the plural pronoun "they"), the practice was certainly understandable. Here are some typical sentences used by all writers in the past and a few today.

> The car hit a groundhog and broke his leg.

> When a person is dieting, he should also exercise.

> If a doctor is consulted, he is likely to advise surgery.

The first example above is not particularly offensive to women, because the "his" refers to an animal. But I am offended, because I think language should be as exact as it can be. If we want to be exact, we should say that a man is a "he," a woman is a "she," and an animal of unknown sex is an "it."

The second example is of a type that abounds in previous literature. The "he" was meant by the writer to include both "he" and "she." With today's increased sensitivities, however, we should discontinue use of such sentences. How? We could change the sentence to "When a person is dieting, *he or she* should also exercise." Such a change is O.K., but, if it is done repeatedly, our sentences tend to become awkward and verbose. Often, the best way to avoid use of the male singular pronoun to include men and women is to pluralize the sentence; then, the plural "they" can be used. In this example, we can avoid the sexism problem by saying "When people are dieting, they should also exercise." Use of the second-person "you" or the third-person "one" are other ways to avoid the male pronouns.

The third example above is not only insensitive; it is factually wrong. When we write such sentences as "the doctor . . . he" and "the nurse . . . she," we are not only being sexist; we are cavalierly overlooking the fact that there are doctors who are "she's" and nurses who are "he's."

Learn to avoid sexist words and phrases. But do so deftly. Don't use strings of "he or she" and "him and her." It is worse to use "he/she" and "him/

her," and the worst of all are such distracting concoctions as "s/he." Some writers have met the pronoun problem by swinging back and forth between "he" and "she," sometimes in alternate chapters. I find such usage terribly jarring and find myself looking back to find nonexistent male or female antecedents.

In my own writing, I find that the best way to avoid sexist pronouns is to use the plural rather than the singular (if possible). Another way is to repeat the noun rather than using a pronoun. Use of "one" is sometimes (but not often) a good answer. Increasingly, I see examples of solving the sexism problem by using a plural pronoun for a singular antecedent. Some people would correct the "doctor . . . he" example above by changing it to "If a doctor is consulted, they are likely to advise surgery." But I see no reason to introduce grammatical errors to correct other errors. I prefer to believe that the best gender-neutral writing is both grammatical and invisible.

## MARITAL STATUS AND SEXUAL ORIENTATION

While the "a doctor . . . they" is fresh in your mind, here is a similar example, taken from a form published by the British Health Service, that introduces my next topic:

> Answer these 3 questions about you, and about your partner if you have one. We use *partner* to mean a person you are married to or a person you live with as if you are married to them.

First, I must say that the "a person . . . them" grates on my ear terribly, even if it does avoid "him or her."

Second, however, I recommend for your consideration this use of the word "partner." The "partners" can be married or unmarried, of opposite sexes, or of the same sex. Isn't "partner" a huge improvement on such terms as "significant other," "housemate," "live-in boyfriend," and "lover"?

So, if the British government can provide health benefits to all people in the U.K. without nosily trying to determine marital status or sexual orientation, why shouldn't all of us extend to each other the same privilege?

## IS MS. A MISTAKE?

Men in our society have always been "Mr.," a title which gives no indication of marital status. Women, considered the chattel of men, were called "Miss" if unmarried and "Mrs." if married. Such labeling made it easy for men to determine which women were "available."

To avoid this semantic labeling of women, the term "Ms." was invented. (Pronounced "Miz," this title was coined by Kansas journalist Roy F. Bailey in 1950.) This title provides a term of address that can be used for both married and unmarried women. Slowly, most style manuals and publishers have come to accept "Ms." as the appropriate title for a woman. Even the staid *The New York Times* now uses "Ms." Therefore, if you want to write with sensitivity, use the title "Ms." in everything you write, from informal letters and memos to journal articles and books. A rare exception might be made for a woman who has made it known that she insists on being addressed in the older terminology. Otherwise, marital status of other people is none of your business. And, by the way, either a "Ms." or a "Mr." with a doctorate is properly referred to as "Dr."

In general, scientific writing and, indeed, all kinds of writing should follow elemental rules of courtesy. Write to inform, not to hurt.

---

Replacing sexist words and phrases with terms that treat all people respectfully can be satisfying and rewarding. It can also be difficult and frustrating, and it is good to admit that.

—Rosalie Maggio

# Appendix 1
# Principles of
# Punctuation Presented
# Plainly

*Everything should be made as simple as possible, but not simpler.*
—Albert Einstein

===========

### Period

Periods are easy to use. Place one neatly at the end of every sentence you write, except questions or exclamations.

This is an easy rule.

### Question Mark

Place a question mark at the end of every question.

Are all rules of punctuation so easy?

Some sentences look like questions, but they may instead be commands. End such sentences with periods. The traffic cop is not asking a question when he or she says:

Would you please pull over.

### Exclamation Point

Put an exclamation point after every real exclamation.

Hold your horses!

As a rule, however, do not use exclamations in scientific writing. Of course, if you discover a new planet or a cure for the common cold, you are entitled to say:

> Eureka!

### Comma

Use a comma after an introductory word (often an adverb) in a sentence.

> Fortunately, this is sound advice.

Use a comma after any introductory phrase.

> On the whole, this is sound advice.
>
> To write well, you should follow this advice.

Use a comma at the end of a dependent clause that precedes the independent clause.

> If you write well, your readers will bless you.

Use a comma and one of the seven coordinating conjunctions (and, or, but, for, nor, so, yet) as one way to separate the two independent clauses in a compound sentence.

> This is a good rule, and your readers will bless you for using it.

Use two commas to set off appositive or interrupting words within a clause or sentence.

> The Director of the laboratory, Dr. Smith, is a good egg.
>
> The Gram stain, named after Alexander Gram Bell, is used in microbiology.

Use a comma before the "and" or "or" in a series.

> Bell also invented other stains, varnishes, and telephones.
>
> Use the right word, phrase, or clause.

Do *not* use a comma between a subject and its verb or between a verb and its object. However, an interrupting word or group of words between a subject and its verb, or between a verb and its object, can be inserted by the use of two (balancing) commas.

> Wrong: I, personally will see to it.
> Wrong: I personally, will see to it.
> Wrong: John Jones, on the first pitch hit a home run.
> Wrong: John Jones on the first pitch, hit a home run.
>  Right: I, personally, will see to it.
>  Right: John Jones, on the first pitch, hit a home run.

### Semicolon

A semicolon can be used to splice together two independent clauses.

Bell invented the phonograph; he also invented the bell.

A semicolon and a coordinating adverb (such words as therefore, however, moreover, etc.) may be used to link together two independent clauses.

This face rings a bell; however, it is not Bell's bell that rings.

Semicolons can be used to separate a series when one or more parts of the series already contains commas. (Note that a semicolon should precede the "and" in the series, just as a comma precedes the "and" in the usual series.)

I gave small bells to Bell, the inventor; middle-size bells to Gram, the painter; and large bells to Ringer, a dead ringer for Bell.

### Colon

A colon is used to introduce a word, phrase, or clause.

Perhaps you have guessed my favorite brand of Scotch: Bells.

I know where to go to get it: to the bar.

When I get it, I might say something like the following: "Hells Bells, and here's mud in your eye."

### Dash

Dashes should be used rarely in scientific writing. Commas or parentheses are usually preferable. A dash might be appropriate if a real contrast or surprise is intended.

My new physician is an odd duck—I have heard that he is a quack.

### Quotation Marks

Periods and commas go inside closing quote marks.

"What is his name," she asked. He said "I am Dr. Quackinabush."

Semicolons and colons go outside closing quote marks.

"Ring the bell"; when he heard those words, he went home.

"Ring it again": those were his words.

Question marks and exclamation points go either outside or inside closing quote marks. They go inside the quotation marks if the quoted person is asking or exclaiming.

> She asked, "May I have a large Bells?"

> "Absolutely!" said the publican.

If it is the narrator who is asking or exclaiming, the question marks or exclamation points go outside.

> Did she say "I want a large Bells"?

> I was amazed when she said "I want a large Bells"!

If *both* the narrator and the person being quoted are asking or exclaiming, the marks go inside.

> Did she say "May I have Absolut vodka, please?"

> His quick response was "Absolutly!"

### Apostrophe

To show possession, add an apostrophe and an "s" to a singluar noun.

> This book is the cat's meow.

To a plural already ending in "s," simply add an apostrophe.

> The scientists' experiments were completed.

To a plural not ending in "s," add an apostrophe and an "s."

> The deer's habits are interesting.

To names, add an apostrophe and an "s."

> Day's Rule
>
> Jones's Rule
>
> Weiss's Rule

### Parentheses

A full sentence within parentheses should start with a capital letter and end with a period; the period should be inside the closing parenthesis mark.

> (I hope you like this rule.)

If the material within parentheses is not a full sentence, any needed comma or period should be placed outside the closing parenthesis mark.

> When you have mastered these rules (punctuational pointers), you will write more confidently.

> When you have mastered these rules, you will write more confidently (as a rule).

### Brackets

Use brackets to enclose any alteration or addition to quoted material.

> "She [the wife of Alexander Graham Bell] was also fond of Bells," he said.

### Slash

Do not use slashes in scientific (or any other) writing, except to indicate division.

> 10/5 = 2

### Hyphen

When two words jointly modify a third, the two should be linked with a hyphen.

> This is a first-rate book.

---

Punctuation rules are important. They are devised to eliminate ambiguities in language. Learn punctuation. Find a handbook such as *The Chicago Manual of Style* and keep it on your writing desk. Few things undercut the authority of a piece of writing more than a simple mistake in punctuation.

—Michael Alley

# Appendix 2
# Problem Words and
# Expressions

*On the most exalted throne in the world, we are still seated on nothing but
our arse.*

—Montaigne (1533-1592)

═══════════════

### *a—an*

Should it be "a history" or "an history," "a hypothesis" or "an hypothesis," "a herb" or "an herb"? It is now correct in the United States (less so
in the U.K.) to use *a* before any word starting with "h" if the "h" is
pronounced, even slightly. Therefore, we say "a history" and "a hypothesis," but of course we add "an herb" to our soup. In the U.S., stay at "a
hotel"; in the U.K., you may stay at "an hotel."

### *access*

In the computer world, *access* is often used as a verb.

I will *access* that file.

In all other uses, however, it is probably better to use *access* as a noun.

May I have *access* to your laboratory?

Some writers, realizing that *access* should be used as a noun, might use
the expression *gain access to;* this expression is acceptable grammatically,
but it is heavy breathing for *enter*.

### affect—effect

The word *affect* is a verb (except in certain psychiatric usage); *effect* is usually a noun.

> A bribe will not *affect* my judgment, unless it is big enough (the bribe, certainly not my judgment).

> Your bribe had an *effect*. (It was big enough.)

Occasionally, *effect* is used as a verb, when its meaning is to bring about or accomplish something.

> Your bribe *effected* a change in my opinion.

### after—following

*After* is the more precise word if a time sequence is involved. If the clowns appear *after* the parade, they will perform after the parade is over. If the clowns appear *following* the parade, they are performing at the tail end of the procession.

### all of the

Say "in all experiments" instead of "in all of the experiments." Similarly, say *both* instead of "both of the," *many* instead of "many of the," etc. Such *of* constructions are not *wrong,* but their repetitive use leads to bloated writing.

### already—all ready

*Already* means *previously,* as in "I have *already* done that experiment." *All ready* means "completely prepared," as in "We are all ready to go on vacation."

### alright—all right

When is it *alright* to use *all right?* It is always *all right* to use *all right.* It is also *all right* to use *alright.* Some grammarians insist that *all right* is more formal (hence, more correct), but I continue to see the use of *alright* by excellent writers. So, if it's *alright* with you, it's *all right* with me to conclude that either is correct.

### among—between

*Between* should be limited to a comparison of two items, and *among* should be used to compare three or more. *Between* you and me, this rule makes good sense; *among* the great unwashed, few people are aware of this distinction.

### amount—number

Use *amount* when you refer to a mass or aggregate; use *number* when countable units are involved. For example, you can say "The *number* of people in our laboratory is 55." You should not say "The amount of people" unless you have weighed them, in which case you can say "The amount of people in our laboratory is about 10,000 pounds."

### and/or

Certainly, this doesn't mean *and* divided by *or*. This slipshod usage can be so confusing that it properly has no place in scientific writing. Theodore Bernstein has described *and/or* as "a visual and mental monstrosity." You should say "A and B," "A or B," "A or B or both," depending on your meaning.

### as (See like—as)

### believe—feel

The short word is usually preferred to the long word. In this pair, however, I *believe* that *feel* is overused. The word *believe* has a clear meaning; *feel* sometimes means "believe," but it often means "touch." If you really *believe* something, say so. If you *feel* something, the reader may assume that you are groping in the dark.

### beside—besides

*Beside* means *near to; besides* usually means *in addition to.*

*Besides* scientific books and journals, I keep mystery novels *beside* my bed.

### between (See among—between)

### both of the (See all of the)

### can—may

These words are often used interchangeably, but confusion sometimes results if you forget that *can* has the connotation of *ability to* and *may* denotes *a grant of permission.* Thus, a graduate student should say "I *can* do that experiment" if the student has the ability to do it. If the lab director has authorized a particular experiment, the student *may* do it.

### case

A *case* of canned goods is fine. A *case* of flu is fine, but you don't feel fine. In most cases, however, the word *case* is used thoughtlessly and repetitively. In some cases, *case* is used to mean "instance" or "example." In other cases, it is not clear what the writer means. In any case, I believe that *case* is the most overused word in scientific writing. In case you are one of the people who repeatedly use *case,* try to eliminate it from your writing, not just in a few cases but in almost all cases.

### compare—contrast

You *compare* two or more items in terms of similarity; you *contrast* differences.

### complementary—complimentary

A *complement* fills up or completes something; a *compliment* is an expression of praise. As yet, I have not had my complement of compliments.

### different from—different than—different to

When in doubt, use *different from.* It is rarely wrong. The word *from* is a preposition, so *different from* is correct whenever *different from* is followed by a prepositional phrase.

A clam is *different from* an oyster.

Because the word *than* is a conjunction, *different than* is correct in those relatively few sentences in which *different than* is followed by a clause.

This experiment posed problems in ways *different than* I had ever imagined.

However, *different than* sentences sound odd (to me at least), even if they are "grammatically correct" (whatever that means). If I were writing the above sentence, I would give the pronoun *from* an object, so that I could say, comfortably and correctly:

> This experiment posed problems in ways *different from* any I had ever imagined.

Do not use *different to* in an American journal.

> This antibiotic is *different to* others produced by *Streptomyces griseus*.

In this example, *different from* is clearly correct in American usage. However, *different to* is accepted in Great Britain. In this instance, British usage is more logical than American. The British can consistently use *different to* and *similar to*, whereas the Americans are stuck with *different from* and *similar to*. But, then, whoever said English grammar is consistent, on either side of the Atlantic?

### disinterested—uninterested

The prefixes *dis* and *un* usually have the same negative meaning. To "disengage a clutch" is the same as to "unengage a clutch." However, if you are interested, "uninterested" means "not interested," whereas "disinterested" means "impartial." An "uninterested" jury is likely to render a poor verdict; a "disinterested" jury is likely to reach a just verdict. Put another way, the disinterested jury will reach a just verdict; the uninterested jury will just reach a verdict.

### effect (See affect—effect)

### epidemic—epizootic

An *epidemic* affects many people in a population; a *pandemic* is the same, except that the "epidemic" occurs over a wide geographic area and affects an exceptionally high proportion of the population. However, some people don't realize that these words come from *demos,* which means people. Thus, we often hear of "epidemics" among horses, rabbits, etc.; that is, we have an epidemic of epidemics, but "epidemics" among animals should be called *epizootics.*

### farther—further

The word *farther* connotes distance, whereas *further* often connotes something additional.

> I hit the ball *farther* than he did.

> Furthermore, I intend to go *further* into this subject.

### *fewer—less*

The word *less* should be used with quantities and qualities; *fewer* should be used with countable units. No matter how many ads you see on television, it is wrong to say that something "has less calories." On the other hand, it is often correct to assume that the advertised product has "less taste." Remember that it takes less effort to do fewer experiments.

### *following (See* **after—following***)*

### *forgo—forego*

*Forgo* means to do without.

> I shall *forgo* the pleasure of your company.

*Forego* means to go before; often, the form *foregoing* is used.

> *Forgo* the spelling "forgoing" in the *foregoing* examples.

### *fraction*

A fraction of scientists carelessly say or write such things as "A fraction of the mice survived." Because "a fraction" can be large or small, the expression is useless. If 1,000 mice were used, "a fraction" of survivors could be anywhere from 1 to 999.

### *further (See* **farther—further***)*

### *imply—infer*

A writer may *imply* something; the reader may *infer* something from what was written. I could *imply* that you are now reading a superb book; from that evidence, you might *infer* that I have gone around the bend.

### *in order to*

This three-word phrase is used relentlessly by every gasbag in the world. All the three words mean is *to*. To learn to write well, learn not to write "In order to learn to write well."

### *in view of the fact that*

Say *because* because (in view of the fact that) one word is better than six.

### irregardless

There is no such word; use *regardless.*

### it should be particularly emphasized that

Such openers are a kind of heavy breathing; they sound exciting, but they say nothing. Omit such preambles and say whatever it is you want to say.

### less (See fewer—less)

### like—as

The word *like* can be used as a preposition; *as* is a conjunction. You should not say "Like I said, I am a botanist." You should say "As I said, I am a botanist."

### many (See much—many)

### many of the (See all of the)

### may (See can—may)

### methodology

Because of the "ology" suffix, methodology means "study of methods." In most scientific papers, you should refer to a *method,* not a *methodology.*

### much—many

*Much* is properly used to describe a quantity or degree, not a number. It is incorrect to say "Maryland led by as much as 34 points." Say "as many as 34 points."

### number (See amount—number)

### perform

An unsuspecting person might think that scientists are monkeys or some other kind of circus animal. They are always *performing.* Some day, I hope that scientists will no longer *perform* experiments. It is much better to do them and be done with it.

### predominate

*Predominate* is a verb and only a verb. *Predominant* is the adjective; *predominately* is the adverb. *Predominate* (which means "to prevail") is a good word, but erroneous usage of these three words predominates.

> I hope that the principles of scientific English will *predominate*. If these principles become *predominant*, scientific writing will improve. Improvement will result because these rules are *predominately* correct.

### previous to (See prior to—subsequent to)

### principle—principal

*Principle* is a noun meaning "a rule of conduct." *Principal* is usually an adjective, meaning "most important." (*Principal* as a noun means a person in charge of a school, or it means a participant in a business deal or some other undertaking.) Remember the principal principle involved here; you must learn to distinguish between these two words.

### prior to—subsequent to

*Prior to* and *previous to* mean *before,* and *before* is shorter and better; *subsequent to* means *after,* nothing more and nothing less. *Before* and *after* are words you should use; *prior to* and *subsequent to* are awful.

### quite

The word *quite* is used frequently in scientific writing, but the word is useless. If you see the word *quite* in any sentence you have written, strike it out and read the sentence again. You will find that, without exception, *quite* is quite unnecessary.

### regardless (See irregardless)

### sacrifice

In certain primitive religious rites, animals or even people were *sacrificed.* In scientific laboratories, however, it is sometimes necessary to *kill* laboratory animals; if it is necessary, please *kill* them and don't euphemistically *sacrifice* them to some god or other. And after they (either the animals or the scientists) are dead, it is because they *died;* they haven't "passed away" or "gone to their reward."

### shall—will

Some grammarians used to go on at great length about the importance of distinguishing between *shall* and *will*. The main part of their game was that *shall* should be used to indicate simple future tense in the first person, and *will* should be used in the second and third persons.

> I *shall* go to Chicago next week.
>
> They *will* go to Chicago next week.

Then, if we wanted to show *determination,* we were supposed to reverse them and use *will* in the first person and *shall* in the second and third.

> I *will* go to Chicago next week.
>
> They *shall* go to Chicago next week.

Well, the grammarians set the rules for this party, but nobody came. People simply do not make these silly distinctions. Would anybody say that Winston Churchill lacked *determination* when he gave his famous "We shall fight on the beaches . . . we shall never surrender" speech? Of course not. Therefore, I say unto you: make no distinction between *shall* and will. Fire at will with either shall or will. And, I *should* add that this same silly distinction once applied to *should* and *would.* So, again, you would be wise to use either *should* or *would* without regard to the pronoun attached.

### should—would (See shall—will)

### subsequent to (See prior to—subsequent to)

### uninterested (See disinterested—uninterested)

### unique

The word *unique* means "having no like or equal." Something is either unique or not unique. If it is unique, it can't be very unique, most unique, or slightly unique.

### utilize

Why use a three-syllable word when a one-syllable word will do? Instead, *use* the one-syllable word. Whether utilizing plumbing facilities or utilizing the most complex scientific instruments, the word *use* is always shorter and always better. Don't use "utilize," and your writing will be much more *use*ful.

### various—varying

If *various* concentrations of a substance were added, they would be defined concentrations (e.g., 10, 15, or 20 mg/ml). However, *varying* concentrations would be undefined and perhaps unmeasured (and changing) concentrations. Vary your language variously to distinguish between these two different words.

### will (See shall—will)

> The pedant is as old as history, but no age has ever taken him so seriously as we do.
>
> —John Jay Chapman

# Appendix 3
# Words and Expressions to Avoid

*If you would be pungent, be brief; for it is with words as with sunbeams. The more they are condensed, the deeper they burn.*

—Robert Southey

| Avoid | Use Instead |
|---|---|
| a considerable amount of | much |
| a considerable number of | many |
| a great number of times | often |
| a majority of | most |
| a number of | some |
| a small number of | a few |
| absolutely essential | essential |
| accompany | go with |
| accounted for by the fact that | because |
| adjacent to | near |
| afford an opportunity | let |
| along the lines of | like |
| an example of this is the fact that | for example |
| an order of magnitude faster | 10 times faster |
| apparent | clear |
| are of the same opinion | agree |
| as a consequence of | because |
| as a matter of fact | in fact (or leave out) |
| as a means of | to |
| as is the case | as happens |
| as of this date | today |
| as to | about (or leave out) |

| Avoid | Use Instead |
|---|---|
| as to whether | whether |
| at a rapid rate | rapidly |
| at an early date | soon |
| at an earlier date | previously |
| at some future time | later |
| at the conclusion of | after |
| at the present time | now |
| at this point in time | now |
| based on the fact that | because |
| be advised that | (leave out) |
| by means of | by, with |
| capability | ability |
| causal factor | cause |
| caveat | warning |
| completely full | full |
| component | part |
| consensus of opinion | consensus |
| considerable amount of | much |
| deem | think |
| definitely proved | proved |
| despite the fact that | although |
| due to the fact that | because |
| during the course of | during, while |
| during the time that | while |
| echelons | levels |
| elucidate | explain |
| employ | use |
| enclosed herewith | enclosed |
| encounter | meet |
| end result | result |
| endeavor | try |
| entirely eliminate | eliminate |
| equivalent | equal |
| eventuate | happen |
| evidenced | showed |
| fabricate | make |
| facilitate | ease, help |
| fatal outcome | death |
| fewer in number | fewer |
| finalize | end |
| first of all | first |
| following | after |
| for a period of | for |
| for the purpose of | for |
| for the reason that | since, because |

| Avoid | Use Instead |
|---|---|
| from the point of view of | for |
| future plans | plans |
| give an account of | describe |
| give consideration to | consider |
| give rise to | cause |
| has been engaged in a study of | has studied |
| has the capability of | can |
| have the appearance of | look like |
| having regard to | about |
| impact (v.) | affect |
| important essentials | essentials |
| in a number of cases | some |
| in a position to | can, may |
| in a satisfactory manner | satisfactorily |
| in a timely manner | promptly |
| in a very real sense | in a sense (or leave out) |
| in almost all instances | nearly always |
| in case | if |
| in close proximity to | close, near |
| in connection with | about, concerning |
| in lieu of | instead of |
| in many cases | often |
| in my opinion it is not an unjustifiable assumption that | I think |
| in order to | to |
| in relation to | toward, to |
| in respect to | about |
| in some cases | sometimes |
| in spite of the fact that | although |
| in terms of | about |
| in the absence of | without |
| in the amount of | for |
| in the event that | if |
| in the first place | first |
| in the not-too-distant future | soon |
| in the possession of | has, have |
| in view of the fact that | because, since |
| inasmuch as | because |
| inception | start |
| incline to the view | think |
| incumbent upon | must |
| initiate | begin, start |
| is defined as | is |
| it goes without saying that I | I |

| Avoid | Use Instead |
|---|---|
| it has been reported by Smith | Smith reported |
| it has long been known that | (I haven't bothered to look up the reference) |
| it is apparent that | apparently |
| it is believed that | I think |
| it is clear that | clearly |
| it is clear that much additional work will be required before a complete understanding | (I don't understand it) |
| it is doubtful that | possibly |
| it is evident that *a* produced *b* | *a* produced *b* |
| it is generally believed | many think |
| it is my understanding that | I understand that |
| it is of interest to note that | (leave out) |
| it is often the case that | often |
| it is recommended that | we recommend |
| it is worth pointing out in this context that | note that |
| it may be that | I think |
| it may, however, be noted that | but |
| it should be noted that | note that (or leave out) |
| it was observed in the course of these experiments that | we observed |
| join together | join |
| lacked the ability to | could not |
| large in size | large |
| let me make one thing perfectly clear | (a snow job is coming) |
| liase with | coordinate with |
| majority of | most |
| make preparations for | prepare |
| make reference to | refer to |
| methodology | method |
| militate against | prohibit |
| month of | (leave out) |
| needless to say | (leave out, and consider leaving out whatever follows it) |
| new initiatives | initiatives |
| not later than | by |
| of great theoretical and practical importance | useful |
| of long standing | old |
| of the opinion that | think that |
| on a daily basis | daily |

| Avoid | Use Instead |
|---|---|
| on account of | because |
| on behalf of | for |
| on no occasion | never |
| on the basis of | by |
| on the grounds that | since, because |
| on the part of | by, among, for |
| optimum | best |
| our attention has been called to the fact that | (we belatedly discovered) |
| owing to the fact that | since, because |
| parameters | limits |
| penultimate | next to the last |
| perform | do |
| permit | let |
| place a major emphasis on | stress |
| pooled together | pooled |
| practicable | practical |
| presents a picture similar to | resembles |
| prior to | before |
| prioritize | rank |
| protein determinations were performed | proteins were determined |
| provided that | if |
| quantify | measure |
| quite | (leave out) |
| quite a large quantity of | much |
| quite unique | unique |
| rather interesting | interesting |
| red in color | red |
| referred to as | called |
| relative to | about |
| remuneration | pay, payment |
| rendered completely inoperative | broken |
| resultant effect | result |
| root cause | cause |
| serious crisis | crisis |
| shortfall | shortage |
| smaller in size | smaller |
| so as to | to |
| subject matter | subject |
| subsequent to | after |
| sufficient | enough |
| take into consideration | consider |
| terminate | end |
| the great majority of | most |
| the opinion is advanced that | I think |
| the predominant number of | most |

| Avoid | Use Instead |
|---|---|
| the question as to whether | whether |
| the reason is because | because |
| the vast majority of | most |
| there is reason to believe | I think |
| this result would seem to indicate | this result indicates |
| through the use of | by, with |
| time period | time, period (not both) |
| to the fullest possible extent | fully |
| transmit | send |
| ultimate | last |
| unanimity of opinion | agreement |
| until such time | until |
| utilize | use |
| validate | confirm |
| very unique | unique |
| was of the opinion that | believed |
| ways and means | ways, means (not both) |
| we have insufficient knowledge | we do not know |
| we wish to thank | we thank |
| what is the explanation of | why |
| with a view to | to |
| with reference to | about |
| with regard to | concerning, about |
| with respect to | about |
| with the possible exception of | except |
| with the result that | so that |
| within the realm of possibility | possible |
| witnessed | saw |

My advice to any young writer is: become an editor. You'll do less work, have less pressure, have more influence, make more money, and best of all: you get to tell others what to do instead of having to do all that rotten research and writing yourself.

—Bob Chieger

# Index

### Compiled by Estella Bradley